2010
To old F

.1 HID RENEWED

MW00844738

More Advance Praise for

The Legacy of Tracy J. Putnam and H. Houston Merritt: Modern Neurology in the United States

"Lewis P. Rowland's personal involvement and meticulous scholarship have preserved the legacy of the distinguished people that spawned the exponential growth of Neurology and Neuroscience in the twentieth century and the great institution whose trainees went forth to lead Neurology departments throughout the country. . . . As we look back, the discovery of Dilantin was the opening salvo of present day translational research, and we are indebted to Lewis Rowland for telling Putnam and Merritt's story and the rise to world prominence of basic and clinical neuroscience investigations at Columbia's Neurological Institute in New York and in the United States. . . . Rowland's book is a great read that may help us effectively meet the challenge of pioneering the future."

—*Stanley H. Appel, MD, Edwards Distinguished ALS Professor, Chair, Department of Neurology, Methodist Neurological Institute, Houston*

"Dr. Lewis Rowland, distinguished neurologist, author, editor, and teacher, has been a major leader in the extraordinary growth and development of Neurology from an arcane specialty to the explosive field of neuroscience over the past fifty years. His scholarship is widely recognized and his encyclopedic knowledge of the history of neurology is reflected in his many publications. Now, with the arrival of his latest book, *The Legacy of Tracy J. Putnam and H. Houston Merritt: Modern Neurology in the United States,* readers interested in the history of neurology will welcome Dr. Rowland's insights into these two icons of twentieth century neurosurgery and neurology. . . . Readers will appreciate his lucid style and his analysis of these events and the individuals who led the way."

—*Robert A. Fishman MD, Department of Neurology, University of California, San Francisco*

"The *Legacy of Tracy J. Putnam and H. Houston Merritt: Modern Neurology in the United States* is a delight to read and will be both exciting and informative to neurologists, psychiatrists, and all with an interest in the history of neuroscience in the twentieth century... In so eloquently telling this story, Bud Rowland, one of Merritt's successors as Chair of Neurology at Columbia's Neurologic Institute, has documented much of the history of modern neurology and neuroscience."

—*John M. Freeman MD, Professor of Neurology and Pediatrics (Emeritus), Johns Hopkins Medical Institutions, Baltimore*

"Dr. Rowland is not only qualified to undertake this project; he is probably the only person alive who could do it. Dr. Rowland trained under Dr. Merritt, and for much of his professional life he has been fascinated by this history. No one else could have tracked down those participants who were around when the story took place. . . . In summary, this is a story that should be told, and if it isn't told now, it never will be."

—*John C. M. Brust, MD, Director, Harlem Hospital Neurology Service, The Neurological Institute of New York*

Left: H. Houston Merritt (1902–1979); *Right:* Tracy J. Putnam
(1894–1975)

The Legacy of Tracy J. Putnam and H. Houston Merritt

Modern Neurology in the United States

Lewis P. Rowland, MD

Professor of Neurology, Chairman Emeritus
Neurological Institute
Columbia University Medical Center
New York, NY

OXFORD
UNIVERSITY PRESS
2009

OXFORD
UNIVERSITY PRESS

Oxford University Press, Inc., publishes works that further
Oxford University's objective of excellence
in research, scholarship, and education.

Oxford   New York
Auckland   Cape Town   Dar es Salaam   Hong Kong   Karachi
Kuala Lumpur   Madrid   Melbourne   Mexico City   Nairobi
New Delhi   Shanghai   Taipei   Toronto

With offices in
Argentina   Austria   Brazil   Chile   Czech Republic   France   Greece
Guatemala   Hungary   Italy   Japan   Poland   Portugal   Singapore
South Korea   Switzerland   Thailand   Turkey   Ukraine   Vietnam

Copyright © 2009 by Lewis P. Rowland.

Published by Oxford University Press, Inc.
198 Madison Avenue, New York, New York 10016
www.oup.com

Library of Congress Cataloging-in-Publication Data

Rowland, Lewis P.
The legacy of Tracy J. Putnam and H. Houston Merritt : modern neurology in the United States /
Lewis P. Rowland.
p. ; cm.   Includes bibliographical references and index.
ISBN: 978-0-19-537952-5
1. Putnam, Tracy Jackson, 1894–1975. 2. Merritt, H. Houston (Hiram Houston), 1902–1979.
3. Neurologists—United States—Biography. 4. Phenytoin—History—20th century.
5. Boston City Hospital. Neurological Unit—History—20th century.
6. Neurological Institute of New York—History—20th century.

[DNLM: 1. Putnam, Tracy Jackson, 1894–1975. 2. Merritt, H. Houston (Hiram Houston), 1902–1979.
3. Neurology—United States—Biography. 4. History, 20th Century—United States.
5. Neurology—history—United States. 6. Neurosciences—history—United States.
7. Phenytoin—history—United States. WZ 100 P993r 2009] I. Title.
RC339.5.R69 2009   616.80092'2–dc22
2008018981

1   3   5   7   9   8   6   4   2
Printed in the United States of America
on acid-free paper

For four generations of Rowlands:
my parents, my wife, my children, my grandchildren

Preface

I never knew Tracy Putnam personally. I knew Houston Merritt from 1950 until he died in 1979.

Together, they discovered the value of a drug called phenytoin (Dilantin) in controlling attacks of epilepsy and thereby helped many thousands of people. They did that great work together on the Harvard Neurological Unit at the Boston City Hospital in the mid-1930s. Eighty years later, the drug is still being used. In 1939, Putnam moved from Harvard to become director of the Neurological Institute in New York, at Columbia. In both places, he was the chief of both Neurology and Neurosurgery, a dual role that proved to be a lethal combination because it guaranteed the enmity of powerful neurosurgeons who did not believe that one person could be both a neurologist and a neurosurgeon. In fact, Putnam was the only one who ever tried to excel in both fields in a major hospital—and he lost, twice.

In the mid-1940s, Putnam was at the peak of his great career. He was then director of both Neurology and Neurosurgery at the Neurological Institute of New York at the Columbia-Presbyterian Medical Center. Nevertheless, he was discharged by the president of Presbyterian Hospital for reasons never really explained. Putnam later said the main reason was his refusal to follow orders from the president of Presbyterian Hospital—that he fire all the Jews on his staff. Putnam moved to California and slowly declined. Two years after Putnam left, Houston Merritt became chair of Neurology at Columbia, and he flourished. So did clinical neurology.

This book is intended to document a unique story about two vital men; the stories include the origins of modern neurology, drug development, neuroscience, and clinical investigation. I have tried to write in a way that not only would be understood by general readers but also would be comfortable for scientists and clinicians.

We owe the book to several individuals. In 1998, I completed my 25-year term as chair of Neurology at Columbia. I planned to write a history of the Neurological Institute of New York, which celebrates its centennial in 2009. However, a few months later, I visited Gerald D. Fischbach, MD, who was then director of the National Institute of Neurological Diseases and Stroke (NINDS). I went as part of a delegation to urge him to put more money into research for Parkinson disease. He cleverly deflected us by asking me what I was doing now that I no longer had administrative responsibilities, so I told him about the planned history.

He said, "I have a better idea. NINDS is having its fiftieth birthday next year. Why don't you write that history?" He quickly made arrangements with Timothy A. Pedley, MD, who had become chair of Neurology at Columbia, for me to spend the year 2000 on sabbatical leave in Bethesda, Maryland. The result was the book *NINDS at 50.*

As chair of Neurology at Columbia, I regularly went to Harlem Hospital to have a clinical conference once a week. When Dr. Pedley became chief, we alternated weeks—and still do. One day, I was walking with John C. M. Brust, MD, director of the Harlem Neurology Service, and he reminded me that the NINDS book had just been finished and asked me if I would now return to the history of the Columbia Institute. I replied that I was no longer up to that job. He quickly and acerbically commented, "You could at least tell the story of Putnam and Merritt." So, more than seven years after that conversation, here it is.

Naturally, many people have contributed generously, including Lawrence K. Altman, Peter Carmel, William D. Caughlin, Ronald Emerson (for advice about restoring vision to blind eyes), Gerald D. Fischbach, Alfred P. Fishman (for information about Isidor Ravdin); Donald H. Harter (for background information about Houston Merritt); Margaret Hogan (for information about the General Education Fund of the Rockefeller Foundation); Russell A. Johnson, Archivist, History & Special Collections Division, Louise M. Darling Biomedical Library, UCLA; Robert Kurtzke, MD (who found the lost tape of an interview with Merritt); Leila Melson, archivist for the American Neurological Association (who provided data for Fred A. Mettler); Michael Myers (for access to his pending history of the Columbia College of Physicians and Surgeons); Stephen Novak, director of Special Collections at the Columbia Medical Library (for information about numerous Columbia people); Timothy A. Pedley; Phillip Person, DDS; Sara Piasecki, Ohio State University Library (for some letters of Tracy J. Putnam); Donald O. Quest, MD (for the history of neurosurgery at Columbia); Thomas Q. Morris, MD; former dean at Columbia College of Physicians and Surgeons and former president of Presbyterian Hospital; Stephen C. Scheiber, former director of the American Board of Psychiatry and Neurology; Collin B. Talley, PhD, historian of Tracy Putnam's interest in multiple sclerosis; Beverly Tetterton, Wilmington (NC) Public Library (for information about the family of Houston Merritt); Lil Wasik, executive administrative assistant, American Board of Psychiatry and Neurology; and Elaine Zimbler, of the Columbia Augustus Long Library (for information about German Jewish refugees at Columbia).

Throughout these years, I have been in constant and repetitive contact with Jack Eckert, reference librarian at the Center for the History of Medicine, Francis A. Countway Library of Medicine, Harvard Medical School. He was always patient, kind, and informative in providing information about Merritt and Putnam at Harvard.

My personal assistant, Hope Poulos, kept records of numerous documents despite my profligate habits and became a reference librarian in the process. She has a keen eye for errors of typography or grammar.

Timothy A. Pedley has given me free rein to divide my time between professorial duties and writing, as well as support in every possible way to complete the book. He and Barbara Koppel, MD, are wise and helpful editorial readers.

Craig Allen Panner, executive editor of neuroscience and neurology at Oxford University Press in New York, made numerous suggestions to improve the text and the figures. Lynda Crawford, Production Editor at Oxford University Press, efficiently guided the process of transforming the manuscript into a real book.

Last but not least are members of my family. The first is Esther Rowland, who has put up with me for 56 years. An accomplished writer and editor herself, she has helped throughout. Our son, Andrew, a professor of epidemiology at the University of New Mexico, is another excellent editor. Our second son, Steve, is a prize-winning radio journalist who taught me the basics of recording interviews. Our daughter, Joy, is a lawyer by day and a sometime writer, too. And her husband, Daryll Alladice, is a middle-school teacher and writer. All have been patient with me and have helped in essential ways.

Lewis P. Rowland, MD
New York City
January 13, 2008

Contents

The Legacy of
Tracy J. Putnam and
H. Houston Merritt

He put water on to soft-boil some eggs and instantly fantasized about having a seizure and pitching onto the stove, charring his face. He stood outside the scene, horrified, but could not seem to call for help . . . he was frightened, and a series of new accidents succeeded the image of his blackened face: a canoe overturned and he convulsed in the water; he fell off a platform and a subway train severed his legs; bus wheels crushed him; he had a seizure while driving and struck an oil truck head-on, exploding his life instantly.

Richard Pollak, *The Episode, A Novel of Suspense.* New York: New American Library, 1986.

1

Dilantin: The Great Achievement of Tracy Jackson Putnam and H. Houston Merritt

*Epilepsy: New Drug Therapy Needed
in the mid-1930s*

Epilepsy comes in many forms, so many different types that specialists have been stymied in trying to come up with a classification agreeable to all of them or, at least, most of them.[1,2] As a result, they are still basing descriptions and discussions on a 1981 document.[3] All epileptic attacks are "brainstorms" that result from disordered electrical rhythms of nerve cells in the brain. The resulting behavioral abnormalities can be seen by others and, too often, by loss of consciousness. If the attack recurs, the condition is called "epilepsy," which is estimated to affect 40 million people worldwide.[4]

Classification of the attacks is important for both diagnosis and treatment, but recognition of the different types now combines not only the visible actions of the subject but also what happens in the electroencephalogram (EEG). The world was simpler for Tracy Putnam and Houston Merritt, when they worked together in the mid-1930s, for the EEG was just then being introduced.

Those of us who do not have epilepsy cannot really know what the patient must feel or fear. Descriptions in novels or portrayals in movies may or may not be accurate, but they seem always to focus on the external manifestations of the attacks—falling, convulsing, tongue-biting, involuntary loss of urine in generalized seizures—or automatic behavior in the "complex partial seizures" of attacks starting in one specific part of the brain, the temporal lobe. Accidental injury and embarrassing behavior are constant concerns to a person with epilepsy. No wonder that patients live with constant worry about the unpredictable next one, a pervasive apprehension that is rarely approached in fictional accounts.

The treatment of epilepsy has been much improved these days, with numerous and effective antiepileptic drugs or therapeutic brain surgery. Combinations of these treatments can totally eliminate the attacks or vastly diminish their frequency and severity. But back in the mid-1930s, before the arrival of Putnam and Merritt, there were only two drugs: bromides and phenobarbital. Both were of limited effectiveness and had serious side effects. Both were sedatives that made people sleepy, and bromides were toxic to the skin as well, causing an ugly and painful rash. The sleepiness caused by these drugs interfered with schoolwork in children and endangered adults who drove cars or sometimes at work.

Epilepsy: The Role of Neurologists

Neurologists are the specialists who take care of people with epilepsy, and there were few doctors in this category in the 1930s because training centers were just beginning to appear. The American Board of Psychiatry and Neurology began certifying neurologists in 1935, but as late as 1947 only 32 residency positions were available in all the United States; by 2006, 451 board certificates were given annually to men and women who had completed the four or more years of training after medical school and then passed written and oral examinations. Growth of the specialty is also documented by the rise of the American Academy of Neurology; from its origin in 1948, membership increased from the 52 charter members to more than 20,000 in 2000.[5]

One of the first neurology training centers in the United States was established in 1930 at the Boston City Hospital under an affiliation with Harvard Medical School. The founder of that division was

Stanley Cobb, and he was interested in epilepsy—as were several other members of that department, especially William Lennox, Fred and Erna Gibbs, and Paul F. A. Hoefer. Husband and wife, Fred and Erna Gibbs were research partners who pioneered in the electrical recording of brain waves, an essential procedure in diagnosing epilepsy. They worked with a Boston electrical engineer, Albert Grass, to develop the first clinically useful devices to record brain waves by EEG.[6] William Lennox later became an international force in improving the lives of people with epilepsy. Tracy Putnam, a young neurologist and neurosurgeon, was trying to establish himself in both fields; formally, he held a research position in the Neurosurgery Division at the Boston City Hospital.

Merritt Joins Putnam at Boston City Hospital: Dilantin

Houston Merritt joined them in 1928 when he enrolled as a resident at the Boston City Hospital for training in neurology. Merritt was 26 years old and seven years younger than Putnam, who was already a Harvard professor. In 1934, Stanley Cobb left the City Hospital and moved back to the Massachusetts General Hospital, another Harvard hospital; Putnam was immediately named to become chief of the Neurology Unit to replace Cobb. With all the interest in epilepsy there, it seems natural for Putnam and Merritt to have started an epilepsy project, too.

Between 1937 and 1941, Putnam and Merritt published seven papers together. Their first publication described a method to determine the seizure threshold of a cat; in doing so, they gave full credit to earlier brain scientists who had used similar methods. With the apparatus they could determine how much electrical current had to be passed through the brain of the animal to elicit an epileptic seizure. Institutional review boards now regulate all experimentation on humans and other animals; no institutional review boards were active in those days to regulate the experiments on animals, which had to be awake during the periods of brain stimulation by electrical current.

In that first paper,[7] they reported that, in contrast to bromides and phenobarbital, the sedative and anticonvulsant effects could be separated. A dose of bromide sufficiently hypnotic to keep the animal from walking raised the electrical threshold by 50%; a dose of phenobarbital sufficient to have the same effect on walking doubled or tripled the convulsive threshold, meaning that after a dose of the barbiturate it took two or three times the electrical current to bring on a seizure. Putnam and Merritt had obtained a series of 19 drugs from the Parke-Davis Company, chosen by chemical structure to resemble phenobarbital. The specific structure was a phenyl group. The first of the 19 drugs they studied was diphenyl hydantoinate; it proved to have the "greatest anticonvulsant activity and the least hypnotic activity."

In their second paper,[8] Merritt's name came first on the list of authors.[8a] They described the effects of many drugs on the seizure threshold in the cat. The primary result was that phenytoin was more effective at raising the seizure threshold and less soporific than phenobarbital.

The third paper[9] appeared in the *Journal of the American Medical Association* (JAMA), again with Merritt as the first author. In the late 1930s and 1940s, JAMA was mailed to almost all American physicians. In this report, they described the beneficial effects of phenytoin in human patients as well as the adverse effects[10]: unsteady gait (ataxia), rash, and gum hypertrophy. None of these problems left lasting damage or was lethal, and all of the adverse effects were reversible. In time, more serious effects were noted, but they were rare. Merritt and Putnam reported some benefit in 87% of all patients tested, including two different seizure types, "grand mal" and "psychic equivalents," which now are called "temporal lobe seizures" or "psychomotor seizures."[11] In the sixth paper,[12] they attempted to identify the chemical structure of the most beneficial drugs, but none was as successful as diphenyl hydantoinate. They continued to write on epilepsy together until 1945, mostly dealing with the same drug.[13-17]

Phenytoin has had several names over the years, but the most popular is the trade name "Dilantin." The generic name changed from "diphenyl hydantoinate" to "diphenyl hydantoin" and then to the current term, "phenytoin."

Whatever the name, the drug has been medically and commercially successful; it has been used throughout the world for more than 70 years, still going strong despite the introduction of many new and competing drugs. Putnam and Merritt seem not to have had a patent, a financial oversight that would be unimaginable for academic scientists in modern times and one that complicated Putnam's later days more than Merritt's, as we will see in Chapter 12. The first patent for Dilantin was USP 2,409,754, issued in October 1946 to Parke-Davis, the pharmaceutical company. That patent expired in October 1963.

Putnam Remembers

Looking solely at their professional publications, one would think that Putnam and Merritt were a great pair of investigators, who should have shared the glow of a glorious achievement. However, in 1945, Putnam became bitter—for good reason—and then turned on Merritt, for a possibly delusional reason, a story we recount later. In 1970, Putnam summed up his view of their joint triumph, bitterly amplifying his own role and denigrating Merritt's[18]:

> In 1923 I began a more intensive training in neurology, as a resident under Dr. Harvey Cushing. He dreaded contact with epileptics who showed no neurological deficits. If one such were by accident admitted to his service, the patient was put at once on Luminal and discharged. ["Luminal" was a trade name for phenobarbital.]
>
> In the late 20s, I began to realize that the cases of epilepsy not susceptible to surgical treatment enormously outnumbered the ones with no localizing signature.[19]

Putnam wrote that he realized biochemical changes could be anticonvulsant, as made evident by the chemical effects of phenobarbital or by changes in the chemical contents of circulating blood brought on by voluntarily overbreathing ("hyperventilation"). These changes reduce the level of carbon dioxide in the blood, which may induce an attack of the petit mal type of epilepsy; conversely, seizures of that type could be stopped by rebreathing into a paper bag or by inhaling carbon dioxide, which would increase blood levels of carbon dioxide and, in so doing, reverse the effects of overbreathing.

Another chemical effect on seizures was the acidosis of the blood that followed prolonged fasting, which William Lennox was studying for antiepileptic effects. When a person fasts, acidic ketone bodies appear in the blood as products of fat metabolism. Lennox himself fasted for two weeks before trying this treatment on a patient.

Putnam continued[19a]: "In 1934 it became clear to me that I should do well to consider the possibility that the medical treatment of various neurological diseases might be improved. Important among these was clearly epilepsy."

Thinking about carbon dioxide, Putnam wondered about placing patients near a brewery, where the environment includes high levels of carbon dioxide, but rejected this as impractical. Then, he thought about phenobarbital. One of Putnam's friends was Tinsley Harrison, an internist and author of a textbook that has gone through many editions and is still a standard reference. Harrison pointed out that convulsions occur commonly in "uremia," a state of chemical changes brought on by kidney failure and characterized by the accumulation of nitrogenous compounds; convulsions, however, are not seen in uremia if the blood also has a high content of chemicals called "conjugated phenols." Therefore, Putnam looked for compounds that contained one or more phenyl groups. He wrote, "The project was brought a step further when I assumed directorship of the Neurological Unit of the Boston City Hospital in 1934."[19a]

At about the same time that Putnam and Merritt were doing research on Dilantin, Fred Gibbs was "setting up the first EEG lab in the world for routine clinical use," also in the Boston City Hospital. Gibbs had shown that seizures are denoted by excessive electrical discharges in the outermost portions of the brain, the cerebral cortex; and he could record abnormal electrical activity on the surface of the skull. Putnam therefore thought about using the convulsive threshold for testing drugs, "quantitating the current required" to cause a seizure before and after administration of the drug being tested. He wrote, "Gibbs, Hoefer, and I assembled a rather crude quantitative stimulator" and Hoefer contributed odds and ends from an electric motor "salvaged from a WWI German airplane." A preliminary trial showed that phenobarbital raised the seizure threshold four times. Putnam continued:

I combed the Eastman Chemical Company's catalog and other price lists for suitable phenyl compounds that were not obviously poisonous. The only one that showed any interest was Parke-Davis. They wrote back to me that they had on hand samples of

19 different compounds analogous to phenobarbital and that I was welcome to them. Among the 19 was diphenylhydantoin. The others were inactive. This one was clearly superior to phenobarbital, nonhypnotic, and apparently nontoxic. Meanwhile a hundred-odd other substances had been tested and a few of them appeared to have some anticonvulsant properties. All except diphenyl hydantoinate had disagreeable side effects and were discarded.

In 1936, I therefore asked Dr. H. Houston Merritt to join us. He was then my faithful Chef de Clinique, in charge of the Out-Patient Department, and as soon as reasonable toxicity experiments had been carried out, it seemed natural to request him to supervise the administration of diphenylhydantoin. He accordingly did so with great success on the very first patient, who became seizure-free.

It was the policy of my department at Harvard to push young men forward. When time for publication came, I suggested to Dr. Merritt that I should be the senior author in publishing the experimental results in the *Archives of Neurology* but that his name should come first on the article on the clinical results. As events turned out, the latter was read by him at the AMA meeting in San Francisco in 1938 and published in the JAMA.

The putdown may not be so clear on first reading, but when work is done by a pair of investigators, the mentor usually expresses pride in the achievements of the protégé. Here, Putnam seems to be claiming credit for everything, magnanimously allowing Merritt's name to come first on the paper describing the human use of the drug. Putnam's brilliance and leadership role are not in question; Merritt's role is challenged.

Dilantin: Merritt's Role

A. J. Glazko,[19b] a ranking officer at Parke-Davis, the company that supplied the drugs, pointed out the error of Putnam's assertion that Merritt was a secondary character, merely pushed forward by his benevolent mentor, and not a full partner in their collaboration. Merritt had been deeply involved in the cat experiments before he took over the clinical trials. Glazko made it clear that corporate leaders were also heavily involved in supplying drugs that were lying on the shelves because they had failed to be effective sleeping pills. After the first batch, the drugs were sent to Merritt for testing anticonvulsant activity.

Merritt's description of the "cat-fit" work was recorded in an interview and published in 1984[20]:

> It was work which Tracy and I did together. Howard (Albert) Grass, who invented the EEG machine, developed an instrument which would deliver a measured amount of induced electric current to the brain. We administered a measured current for an arbitrary two minutes and ten seconds. By that time we could tell whether a cat was going to have a fit or wasn't. Then we could run the rheostat up, so that the cat would get a larger amount of current next time. In this way we could find out the amount of current necessary to produce a convulsion in the animal. Then we tried a number of drugs, hoping that they would prevent the convulsions. We started with drugs that were known to have an analeptic effect, such as phenobarbital and other hypnotic drugs. Then we tried others that were not known to be hypnotic. We asked the pharmaceutical houses to send us some drugs which had been devised as hypnotics but had not been useful in that capacity. Parke-Davis was one of the first drug companies we asked, because the Director of Research there was a friend of Tracy Putnam. Parke-Davis sent about twenty-five drugs that had not worked as hypnotics. The first drug which they sent was diphenylhydantoin, and, by God, it worked. We tried about a thousand drugs after that. We found one or two which were more effective anticonvulsants than diphenylhydantoin, but they were too toxic to use.

Dilantin: The Legacy

Putnam died in 1975 at age 81; Merritt died in 1979 at age 77. In 1982, 55 years after the discovery, Parke-Davis instituted the Merritt-

Putnam Symposia. In the preface to the publication of papers[21] from that first meeting, I credited them with the following achievements:

1. They devised a simple and reliable method to test drugs for anticonvulsant effect.
2. They showed that anticonvulsant effects in cats accurately predicted effects in humans; phenytoin was the first anticonvulsant drug to be tested in animals before it was given to human subjects.
3. They showed that anticonvulsant and sedative effects of drugs could be separated.
4. They showed that a single drug might be much more effective in treating some seizure types than others. This not only had therapeutic implications for choice of drugs to treat individual patients; it also implied that the pathogenic mechanisms of different seizure types were also different.
5. They discovered the efficacy of phenytoin and it was being used in thousands of patients.
6. Their work opened the way to the development of other anticonvulsant drugs.

I could have added a seventh effect. Theirs was one of the first studies that would now be called "translational research," the application of basic science results in laboratory animals to a disease in humans, with an effective treatment. In that way, their experiments were among the first that warranted the term "clinical investigation."

However, five years after the Putnam-Merritt Symposium, Walter Friedlander[22] waylaid that summary as inaccurate "common wisdom." He complained that others had studied anticonvulsants in animals with seizures induced by chemicals. It was also known already that ketones, which appear in the blood after periods of starvation, could ameliorate attacks induced chemically—without sedation. He noted that others had tested drugs in animals before giving them to humans but that the "drugs" were called "vital dyes" and came to naught as a human therapy. Friedlander wondered why Putnam and Merritt relied on electrical stimulation to induce seizures when chemicals could also do it. He described earlier attempts to induce seizures electrically, as Putnam and Merritt had explicitly acknowledged in their first paper.

Friedlander's paper may still be the most comprehensive review of the historical background for the discovery of phenytoin. Whatever his reservations about crediting Merritt and Putnam, he acknowledged that their work led to what he himself called "a paradigm shift" in recognizing that new drugs could be developed, identified, characterized, tested in the laboratory, and then applied to people. Friedlander lauded the flurry of new anticonvulsants that followed Dilantin. In the 37 years from 1937 to 1974, 17 new drugs appeared (in contrast to the two that had preceded phenytoin in the 81 years from 1857 to 1938). Ten new drugs appeared after 1990, and still they come.

Some authorities think this scientific advance warranted a Nobel Prize, but Merritt and Putnam were never nominated (according to the Nobel website). Others have lauded the achievement. For instance, Benjamin White wrote in 1984[23]

> . . . the introduction of diphenylhydantoin (Dilantin) was the greatest advance in treatment of epilepsy since the introduction of phenobarbital by Hauptmann in 1912. Dilantin was the first non-sedative drug to be useful, and it remains a cornerstone of therapy even to this day.

More simply but emphatically, Eli Goldensohn, a neurologist, leader of epilepsy research, and editor of the journal *Epilepsia*, claimed that their work "improved the lives of millions of people" and their papers "continue to exert a profound influence on investigations into the causes and treatments of epilepsy."[23a, 23b]

Gibbs and Lennox won the Lasker Award for Clinical Medical Research in 1951; Merritt and Putnam were ignored.

Anthony Glazko, the officer at Parke-Davis, clarified some of the history.[24] Putnam had written to Oliver Kamm, then scientific director of the company. W. Glen Bywater selected the compounds from the company's collection for anticonvulsant testing and his colleague, Arthur W. Dox, wrote to Putnam when the samples were ready; they were sent in April 1934. Bywater became the contact and he "communicated mainly with Merritt." Glazko wrote

> . . . the screening procedure had not been fully established at the time the first samples were received, and further work was required.

Merritt must have been heavily involved in the development of the screening procedure, although Putnam seemed to avoid any mention of Merritt's role in the animal work. However, Merritt was a co-author in the very first paper describing the screening procedure. He was also the senior author in the meeting presentations and in the publications reporting the animal work, as well as the clinical observations.[24]

Posterity has been kind. In 1984, JAMA reprinted the 1938 paper as a "landmark article."[25,26] The contributions of Putnam and Merritt were summarized by Tyler et al.[27]: "Their discovery ranked as the first major successful medical treatment for epilepsy in the 20th century." In writing on the history of epilepsy therapy, D. F. Scott called phenytoin "a major advance."[28]

How Putnam split from Merritt is a sad, sad story, one that erupted a decade after their historic publication.

Notes

1. Engel J Jr, Williamson PD, Berg AT, Wolf P. Classification of epileptic seizures. In Engle J Jr, Pedley TA, eds. Epilepsy, a comprehensive textbook, 2nd ed, vol 1. Philadelphia: Williams & Wilkins, 2008, pp 511–519.
2. Kellinghaus C, Lüders H, Wylie E. Classification of seizures. In Wylie E, ed. The treatment of epilepsy. Principles and practice, 4th ed. Philadelphia: Lippincott Williams & Wilkins, 2006, pp 217–229.
3. Commission on Classification and Terminology of the International League Against Epilepsy. Proposal for revised clinical and electroencephalographic classification of epileptic seizures. Epilepsia 1981;22:489–501.
4. Bazil CW, Morrell M, Pedley TA. Epilepsy. In Rowland LP, ed. Merritt's neurology, 11th ed. Philadelphia: Lippincott, Williams & Wilkins, 2005, pp 988–1012.
5. Rowland LP. NINDS at 50. New York: Demos Medical Publishing, 2003, p 124.
6. Zottoli SJ. The origins of The Grass Foundation. Biol Bull 2001; 201(2):218–226.
7. Putnam TJ, Merritt HH. Experimental determination of the anticonvulsant properties of some phenyl derivatives. Science 1937;85:525–526.
8. Merritt HH, Putnam TJ. A new series of anticonvulsant drugs tested by experiments in animals. Arch Neurol Psychiatry 1938;39:1003–1015.
8a. Being the first author is important because that name is the one used in future citations of the work. It denotes who did most of the scientific work and, usually, who wrote the first draft of the article. The senior member of a

multi-author book is usually listed last. Other contributors are listed in the middle. In many journals, only the first three authors are named and the others are called "et al."

9. Merritt HH, Putnam TJ. Sodium diphenyl hydantoinate treatment of convulsive seizures. JAMA 1938;111:1068–1073.

10. Merritt HH, Putnam TJ. Sodium diphenyl hydantoinate treatment of convulsive seizures. Toxic symptoms and their prevention. Arch Neurol Psychiatry 1939;42:1053–1058.

11. Merritt HH, Putnam TJ. Further experiences with the use of sodium diphenyl hydantoinate in the treatment of convulsive seizures. Am J Psychiatry 1940;96:1023–1027.

12. Putnam TJ, Merritt HH. Chemistry of anticonvulsant drugs. Arch Neurol Psychiatry 1941;45:505–516.

13. Merritt and Putnam, Further experiences.

14. Merritt HH, Putnam TJ. Experimental determination of anticonvulsive activity of chemical compounds. Epilepsia 1945;3:51–75.

15. Merritt HH, Putnam TJ, Bywater WG. Synthetic anticonvulsants; 5,5-disubstituted hydantoins containing hetero-atom in side chain. J Pharmacol Exp Ther 1945;84:67–73.

16. Merritt HH, Putnam TJ, Bywater WG. Anticonvulsant sulfoxides and sulfones. Arch Neurol Psychiatry 1945;54:319–322.

17. Putnam TJ, Merritt HH. Dulness (sic) as an epileptic equivalent. Arch Neurol Psychiatry 1941;45:797–813.

18. Putnam TJ. The demonstration of the specific anticonvulsant action of diphenylhydantoin and related compounds. In Ayd FJ Jr, ed. Discoveries in biological psychiatry. Philadelphia: Lippincott, 1970, pp 85–90.

19. Scars in the brain or tumors in the brain can cause seizures, and the manifestations of the seizure may be clues to the location of the scar in the brain. For instance, an attack might start with shaking of one hand, and the scar would be in the "hand area of the motor cortex." Lesions elsewhere in the brain might disturb speech or vision. Clinical deductions can be validated by modern brain imaging, and removing the scar tissue surgically may ameliorate the seizures. Cushing may not have been interested in surgery for epilepsy, but Wilder Penfield pioneered that field when he left the New York Neurological Institute for the one in Montreal, as described in Chapter 3.

19a. Putnam and Merritt, Dulness (sic) as an epileptic equivalent.

19b. Glazko AJ. Historical commentary. Discovery of phenytoin. Ther Drug Monitor 1986;8:490–497.

20. White B. Stanley Cobb: A builder of modern neuroscience. Boston: Francis A. Countway Library, 1984, p 194.

21. Rowland LP. Introduction to the Merritt-Putnam symposium. Epilepsia 1982;23(suppl 1);S1-S4.

22. Friedlander W. Putnam, Merritt, and the discovery of Dilantin. Epilepsia 1987;28:S1-S21.

23. Recorded interview with H. H. Merritt, date uncertain but after arrival in New York in 1975. Archives and Special Collections at Columbia University Health Sciences Library.

23a. Personal communication, August 2003.

23b. Goldensohn ES. Merritt-Putnam: The Legacy. Epilepsia 1992;33 (Suppl 4): S3-S5.

24. Glazko AJ. Addendum to "Putnam, Merritt, and the discovery of Dilantin." Epilepsia 1987;28:87–88.

25. Merritt HH, Putnam TJ. Landmark article Sept 17, 1938. Sodium diphenyl hydantoinate in the treatment of convulsive disorders. By H. Houston Merritt and Tracy J. Putnam. JAMA 1984;251:1062–1067.

26. Van Allen MW. Landmark perspective. Introduction of sodium diphenyl hydantoinate. JAMA 1984;251:1068–1069.

27. Tyler K, York GK, Steinberg DA, et al. Part 2. History of 20th century neurology: decade by decade. Ann Neurol 2003;53(suppl 4):S27-S45.

28. Scott DF. The history of epileptic therapy: an account of how medication was developed. Carnforth, UK: Parthenon Publishing Group, 1993

There is, however, in New England, an aristocracy, if you choose to call it so, which has a far greater character of permanence. It has grown to be a caste, not in any odious sense, but, by the repetition of the same influences, generation after generation, it has acquired a distinct organization and physiognomy, which not to recognize is and not to be willing to describe would show a distrust of the good sense and intelligence of our readers, who would like to have us see all we can and all we see.

Oliver Wendell Holmes, Sr. The Brahmin caste of New England. In the series "The Professor's Story." *Atlantic Monthly* 1860;5(27):92–93.

2

Putnam, a True
Boston Brahmin

Tracy J. Putnam, Neurobiologist Extraordinaire:
Being a Brahmin

We know little about Houston Merritt's origins, but his professional life is documented from the time of his arrival in Boston in 1928 to his prime years, his final illness, and even his postmortem examination in 1979 (as described in Chapter 13).

In contrast, Tracy Putnam's family is thoroughly documented for two centuries, but little is known about his final years.

Putnam was the scion of five European families that were among the first to appear in North America: the Jackson, Quincy, and Pickering families, who settled Massachusetts Bay Colony, as well as the Washburn and Tucker families, of the Plymouth Colony. He was therefore a true Boston Brahmin, as defined by Oliver Wendell Holmes, Sr., the physician–writer and father of the even more famous Supreme Court justice who had the same name plus "Junior"; they were Brahmins themselves.

Holmes[1] is often credited with originating the term "Boston Brahmin," but he actually used the term "Brahmin caste of New England," so someone else later converted the locale and the term to Boston. According to the *Oxford English Dictionary* (OED), an essay about

Holmes came close again in 1891 but did not actually use what William Safire would call the "wedded words"[2]:

> To be a missionary of Boston culture, rather than an apostle of political or theological revolution, must have pleased the anxious thought of this medical Brahmin. For Boston was then, and has ever since been to him, the chief city of the world.

The next OED citation could have been as late as 1931, when James Truslow Adams wrote the following[3]:

> It was the West that was building up Eastern manufacturing and Eastern fortunes, and it was the West that was dominating the American mind and outlook, in spite of the smug Boston Brahmins, and shipowners, New York bankers, or Southern cotton magnates.

By analogy with the Brahmins of India, Holmes implied that they were a group privileged by genealogy, wealth, education, scholarship, and public achievement. For almost 200 years they dominated commerce, politics, literature, politics, and Harvard University. Recognizing the privileges of inheritance, they were generally concerned with all of society, including the poor and underprivileged.[4] William Lloyd Garrison was joined by many other Brahmins in opposing slavery. Among the typical Brahmin characteristics, descendants married each other in a royal tribal sequence, and so did Tracy Putnam. His wife descended from the Brewster and DeWolfe families of the Plymouth Colony.

Four Generations of Josiah Quincy

According to a curriculum vitae prepared after 1947, Putnam wrote that one of his forebears was Josiah Quincy, the name of men in four generations of the family (without being identified by "Junior" or a Roman numeral). The first Josiah Quincy (1709–1784) was a Revolutionary War soldier. His son (1744–1775), a lawyer, was also an American Revolutionary War leader who wrote anonymous newspaper articles for the *Boston Gazette* opposing the Stamp Act and traveled to England to protest the closing of Boston Harbor after the Boston Tea Party and other grievances against the British government. He died en route home from London. Despite his revolutionary role, he partnered with John Adams to defend British soldiers after the 1770 Boston Massacre, in which five civilians were killed by British troops.

His son, the third Josiah Quincy (1772–1864) became the second mayor of Boston (1823–1828), served in Congress as a U.S. representative (1805–1813), and was president of Harvard (1829–1845).[5] The fourth Josiah Quincy (1802–1882) was also a Boston mayor (1846–1848). The town of Quincy is named for them, and Quincy House, built in 1770, is a national monument.[6]

The Putnam Lineage

Another early Putnam was Samuel, a judge on the Massachusetts State Supreme Court and brother of Israel Putnam, who was one of George Washington's favorite generals in the revolutionary army. At Harvard, the Putnam family contributed James Jackson, the first Hersey Professor of Physick; James Jackson Putnam, first professor of neurology; and George Minot, professor of medicine and 1934 Nobel laureate for discovering the role of the liver in curing pernicious anemia.

But Brahmins were not always angelic. Members of the family played a role in the witchcraft trials in Salem,[7] an episode later dramatized in *The Crucible* by the great American playwright Arthur Miller. Thomas Putnam "took a most prominent part in the witchcraft delusion of 1692, being in fact second to none but Rev. Parris in the fury with which he seemed to ferret out the victims of his young daughter's insane for notoriety."[8]

About Mrs. Thomas Putnam, it was said that "at the trial of the accused persons, she was often seized with strange attacks of imaginations, evidently produced by overexcitement and the strain on her brain. At these times she was a prominent witness."[9] Another woman who claimed to have been possessed was Ann Putnam.[9a] Deacon Edward Putnam was a member of the group that brought charges against "so many innocent people."[10]

On the other side was Joseph Putnam, who opposed the proceedings "from first to last." He recognized his peril in taking this stance "and for six months, one of his fleetest horses was kept saddled, ready at a moment's notice should an attempt be made to seize his person."[11]

On both sides was Jonathan Putnam. He and Deacon Edward

were the complainants for the warrant issued against Rebecca Nurse and Dorcas Good, the latter a child of but four or five years of age. Afterward, however, Jonathan Putnam saw his mistake and with characteristic manliness signed the paper declaring that in his belief she could not be guilty of the charge preferred against her.[12]

James Jackson Putnam, Pioneer Neurologist

Putnam's father was a physician, but his brother, Tracy's uncle, became more noteworthy. James Jackson Putnam was born in 1846 and died in 1918; he became the first professor of neurological disease at Harvard, was one of the seven founders of the American Neurological Association, and was president of that organization in 1888. He became interested in psychoanalysis and was the one who invited Sigmund Freud to give the five 1909 lectures at Clark University, which resulted in an influential book and boosted Freud's international fame. Harvard established the James Jackson Putnam Professorship in Neurology, which was probably the very first endowed chair in the field and in the United States.

The Medical Education of Tracy J. Putnam

Tracy Putnam interrupted his undergraduate years to serve as an ambulance driver in France during World War I and was honored by the Croix de Guerre in 1915. He received his AB degree from Harvard in 1916 and his MD in 1920. According to his curriculum vitae, he had the highest grades in his class. As a medical student, he published one paper on occupational toxicology in workers exposed to trinitrotoluene, another paper on the effects of salt ingestion on the pressure of cerebrospinal fluid, and another on the area postrema, a tiny brain structure swathed in ignorance as well as being fully exposed to the cerebrospinal fluid. He also published a paper on a condition called "hydrocephalus," a disorder of the circulation of cerebrospinal fluid that can be devastating to infants. His research career had begun with flair, especially for a medical student.

On graduating, he was a resident in the Department of Pathology at Johns Hopkins in 1920–1921; there, he met and later married Dr. Irmarita Keller, a developmental psychiatrist who was psychoanalyzed by Sigmund Freud himself. Putnam returned to Boston in 1921 as a surgical house officer at Massachusetts General Hospital until 1923. Then, for the year 1923–1924, he moved to the Peter Bent Brigham Hospital for training in neurosurgery with Harvey Cushing, the founder of modern neurosurgery in this country. With Cushing, Putnam published a description of the pathology of subdural hematoma, a blood clot on the external surface of the brain that arises from head trauma and can usually be cured by a simple operation.

It was customary at the time for aspiring academic physicians to go to Europe for training in research. Most went to Germany, but in 1924 Putnam went to Amsterdam, where he studied with Barends Brouwer, an authority on the microanatomy of brain regions involved in vision. There, in his own words, he "carried out an extensive investigation of the anatomy of the visual system in animals" and humans.[13]

Putnam's choice was emblematic of his wide-ranging, eclectic approach to research. Looking back, he did not act like a conventional neurosurgeon. Cushing was the leader of the first generation. He was interested in operating on brain tumors—as was Earnest Sachs in St. Louis. Others in that first generation also thought in terms of tumors. Walter Dandy, at Johns Hopkins, made a notable contribution by leading the way to imaging brain tumors by injecting air into the cerebrospinal fluid that fills the spaces of the brain; mass lesions distorted the normal pattern as seen by X-ray examination and, thus, could be diagnosed (albeit much more crudely than today's hi-tech computerized tomography and magnetic resonance imaging). Also in that first generation was Charles Elsberg at the Neurological Institute of New York; he pioneered in spinal cord tumors and other mass lesions that could be removed surgically.

Putnam surely diverged from that pattern. In 1925, after a year in Amsterdam, he returned to Boston as a fellow in charge of a laboratory of surgical research. Also starting in 1925 and continuing until 1928, he was a neurosurgical resident with Harvey Cushing at the

Peter Bent Brigham Hospital. By then, he must have been a true neurosurgeon. Although he had Harvard academic titles, he seems to have done little at the Massachusetts General Hospital itself, and he was barely mentioned in a history of the neurosurgical service there.[14]

Nevertheless, by 1930 Putnam had proved to be both a neurologist and a neurosurgeon, doing research in both fields. No one, before or after Tracy Putnam, had ever achieved premier recognition in both areas of endeavor. This is not mere assertion because his credentials were documented later. In neurology, he was to become chief editor of the *Archives of Neurology and Psychiatry*. He also carried the neurosurgical cachet of having been trained by Cushing. Putnam was one of the original founders and the first secretary-treasurer of the Harvey Cushing Society of neurosurgeons, founded in 1932; and later, when that organization changed its name to the American Association of Neurological Surgeons, Putnam was president.

Putnam was surely unique, and in 1930 he was ready for the next chapter in his life story, when he moved to the Boston City Hospital.

Notes

1. Holmes OW Sr. The Brahmin caste of New England. In the series The professor's story. Atlantic Monthly 1860;5(27):92–93.
2. Sanborn FB. Oliver Wendell Holmes. In The Homes and Haunts of Our Elder Poets. New York: D. Appleton and Company, 1891, p 155.
3. Adams JT. The epic of America. Garden City, NY: Garden City Publishing, 1931, p 206.
4. O'Connor TH. Bibles, Brahmins and bosses. A short history of Boston, 3rd ed. Boston: Trustees of the Boston Public Library, 1991.
5. Mann A. Josiah Quincy, www25.uua.org/uuhs/duub/articles/josiahquincy. html
6. Viera D. Hidden spaces of old houses conceal clues and curiosities. Historic New England Magazine 2001(Fall), p 2.
7. Descendants of John Putnam, http://wwww.billputnam.com, pp 1–80.
8. Adams, The epic of America, 8.
9. Ibid., 9.
9a. Ibid.
10. Ibid., 11.
11. Ibid., 15.
12. Ibid., 24.
13. Research on vision is described in a typescript entitled "Neurosurgical Research of Tracy J. Putnam, M.D." It mentions research done as late as 1953

and was therefore written when he was already in Los Angeles. It is included in his papers at the UCLA Library.

14. Zervas NT, ed. Neurosurgery at the Massachusetts General Hospital—1909 to 1983—A short history and alumni record, 1st ed. Boston: Massachusetts General Hospital, 1984; http://www.billputman.com/THE%20NEW%20E NGLAND%20PUTNAM%20FAMILY.htm

3

Boston City Hospital: Cradle of Modern Neurology in the United States

The University of Pennsylvania School of Medicine was the first and only medical school in the 13 American colonies when, in the fall of 1765, students enrolled for "anatomical lectures" and a course on "the theory and practice of physick." They enrolled at the College of Philadelphia, which was the name of the University of Pennsylvania in pre-Revolutionary times.[1]

King's College organized a medical faculty in 1767 and was the first institution in the North American colonies to confer the degree of doctor of medicine. The first graduates in medicine from the college were Robert Tucker and Samuel Kissarn, who received the degree of bachelor of medicine in May 1769 and that of doctor of medicine in May 1770 and May 1771, respectively. Instruction in medicine was given until interrupted by the Revolution and the occupation of New York by the British, which lasted until November 25, 1783. In 1784 instruction was resumed in the academic departments, and in December of the same year the medical faculty was reestablished. In 1814 the medical faculty of Columbia College was merged with the College of Physicians and Surgeons.[2]

On September 19, 1782, the president and fellows of Harvard College adopted a report, presented by President Joseph Willard, embodying plans for a medical school. With a handful of students and a faculty of three, classes at the Medical School began in Harvard Hall in the college yard and later were transferred to Holden Hall, originally the college chapel.[3]

> Dean Edsall visited me and said he wanted to develop a department of neurology at Harvard. "I was thoroughly sympathetic." He had a well worked-out plan. I replied, "Send the right man abroad to study and let him on his return to America gradually build up his clinic and his staff. Moreover, you are lucky, for you have the man in Stanley Cobb."[4]

Philadelphia, New York, and Boston vie for the honor of pioneering medical education in the United States. To some extent, claims for the priority of each of these medical schools also included neurology. Pennsylvania had notable neurologists in the nineteenth century, especially S. Weir Mitchell, who was confronted with battle wounds involving the nerves of arms and legs during the Civil War. He founded a neurological hospital afterward and became famous as a novelist. Later, in the twentieth century, C. J. Mills and W. G. Spiller were prominent active neurologists at hospitals affiliated with the University of Pennsylvania in Philadelphia. But their legacy was limited because the few neurologists they trained tended to practice locally; they were Philadelphians, and after training, they stayed in the neighborhood and had a limited influence nationally.

A little farther north, the New York Neurological Institute was founded in 1909 as the first hospital in North America for people with diseases of the nervous system. It became important to the story of Putnam and Merritt, but during the 1930s, the Boston City Hospital dominated the evolution of neurology. We have already seen how epilepsy research flowered there (Chapter 1). The memorable names are Stanley Cobb, Hallowell Davis, William Lennox, Fred and Erna Gibbs, Frank Fremont-Smith, and then Putnam and Merritt. How did that come about?

Stanley Cobb

The central figure at the Boston City Hospital was Stanley Cobb, born in 1887 to a wealthy Boston business family. They qualified for Brahmin status because Thomas Cobb came to the area with the founders of the city. He was born "about 1762" and died "before 1848."[5]

Stanley Cobb was a shifted sinistral, born left-handed and taught to use his right hand. Although left-handedness was once thought to predispose a person to stuttering, that theory has long been supplanted by genetic susceptibility. Regardless of theory, the speech disorder seriously embarrassed Stanley, so much so that his mother kept him at home and out of school until he was eight years old. Like many other stutterers, he had a superior intellect; he attended Harvard College and then Harvard Medical School in the class of 1914. In medical school, he befriended David Edsall, then the Jackson Professor of Medicine. Cobb married Elizabeth Almy in 1915, and they moved to Baltimore, where he did research in physiology at Johns Hopkins. He returned to Boston in 1919 to continue his research in physiology with two great Harvard experimenters, Walter B. Cannon and Alexander Forbes, who were stationed at the medical school quadrangle. In 1923, Cobb, Elizabeth, and their children went to Europe and toured. There, he met with leading neuroscientists and became fluent in both German and French.[6]

In 1923 also, David Edsall became the first full-time Dean at Harvard Medical School. He wanted to create a department of neurology that would build on the fame of James Jackson Putnam. The Department of Medicine had developed a group at the Boston City Hospital, which seemed a solid enough place for neurology, too. Edsall therefore sought both advice and money from Abraham Flexner, the famous critic of American medical education and champion of full-time faculties (who would teach and do research for a salary rather than depending on a private practice that might distract the physicians from their primary goals). His influence was fortified because he was the dispenser of funds from the General Education Board of the Rockefeller Foundation.

In his practice, Cobb had treated a man with a serious depression, which fortunately cleared without modern drug therapy. The patient happened to be a friend of Flexner's, who thereby became yet another admirer of Cobb. Through the Education Board, Flexner would donate $350,000 in 1925 to facilitate the neurology program at the City Hospital when Cobb returned from his two-year tour in Europe. The sum was to be matched by Harvard. The last piece in the project was the construction of a new building at the hospital, but that was delayed for another five years.

In the meantime, illustrious colleagues worked in Cobb's neuro-pathology laboratory on the medical school campus. Among them were the epilepsy group including Lennox and Fremont-Smith, as well as others: Henry Forbes, Harold G. Wolff, Paul Yakovlev, John Talbott, Raymond Morrison, Jacob Finesinger, George Schaltenbrandt, and Samuel Epstein. Cobb, now relieved of private practice and on a full-time salary, managed to perform research and publish papers with all of them individually or in small groups. They all became distinguished, but Cobb regarded Wolff as "the most valuable assistant he ever had,"[6a] especially in their work on the cerebral circulation. Wolff later became the founder of the neurology department at the Cornell-New York Hospital Medical Center.

In 1928, Cobb went to Europe again and Fremont-Smith filled in temporarily, aided by Edward Gildea. Tracy Putnam, "a neurosurgeon of academic stature,"[6b] was brought on staff to help neurosurgeon Donald Munro. Houston Merritt came in 1928 as a trainee in neurology. In 1930, the new hospital building finally opened at the Boston City Hospital and the investigators moved as a group. In one way or another, almost all worked on something related to epilepsy, including Putnam and Merritt.

Four years later, in 1934, Cobb moved to the Massachusetts General Hospital as chief of Psychiatry. That was the first department of psychiatry in an American general hospital, setting the model for other hospitals and establishing Cobb's role as a leader of psychiatry, combining a biological approach and psychoanalysis. Putnam immediately succeeded Cobb as head of Neurology at the City Hospital, and Houston Merritt matured.

Frank Fremont-Smith, Raymond D. Adams, and Others

Merritt worked intimately with Putnam in establishing the value of Dilantin and did more of his own separate research. He collaborated with Frank Fremont-Smith on the constituents of cerebrospinal fluid (CSF). Fremont-Smith had started that research, but others considered him unreliable. For instance, he had been publishing regularly with Cobb but left for a European journey without completing his part of one project. As a result, Cobb published nothing with him for the next six years. Nevertheless, Merritt managed to partner with

Fremont-Smith, and their 1937 book on the constituents of CSF in health and disease became a classic.

Merritt also studied neurosyphilis. That effort resulted in another monograph, this one written with psychiatrist Harry Solomon and neuropathologist Raymond D. Adams. During World War II, Merritt became an expert on head injury, too.

Putnam later considered Merritt a mere assistant in the discovery of Dilantin, and some others shared that view.[7] Nevertheless, as we saw in Chapter 1, Merritt ultimately received high praise for that achievement. Merritt was also universally admired and appreciated by those around him because of his affability and his remarkable clinical skills. He would see patients in consultation at any time. He made rounds at the City Hospital on Sundays and would then go to other Harvard hospitals to see patients who were diagnostic or therapeutic challenges.

Fred and Erna Gibbs, Hallowell Davis, and William G. Lennox

In 1939, Putnam left to become director of the Neurological Institute in New York. Merritt became the interim head at Boston City Hospital and was the local favorite to become Chief of Neurology, but that was not to be. Instead, the search committee picked a New Zealander working in England, Derek E. Denny-Brown. He had been doing research with Nobel laureate neurophysiologist Sir Charles Sherrington.

The epilepsy legacy of the Neurological Unit at Boston City Hospital was literally fabulous, almost unbelievable. Not only were the investigators remarkable but also there was no other research center remotely as productive. When Fred Gibbs graduated from Johns Hopkins Medical School in 1929, he "decided to study epilepsy and the only place I knew where epilepsy research was occurring was at Harvard with Lennox and Cobb."[8]

The story may have started with Hallowell Davis, who began to do research at Harvard in 1922 and was one of the first to use electrical records of brain waves.[9] Both "electrical" and "records" were new advances in epilepsy research. Davis continued to collaborate with Lennox and Gibbs and later became the world's premier authority on the physiology of hearing, publishing scientific papers for 80 years, some in conjunction with his wife, Pauline. They met and married in 1923, while they were members of a medical relief team in Istanbul, carrying out one of his many societal contributions. He moved from Harvard after 24 years without ever being promoted from the rank of instructor, the lowest rung on the academic ladder. However, he then became director of research at the Central Institute for the Deaf, an affiliate of Washington University in St. Louis, where he was given the title of associate professor on arrival.

While still in Boston, Davis was the one who identified Albert M. Grass, a 1934 engineering graduate of the Massachusetts Institute of Technology, as the person to build the equipment for what they called "electroencephalography," or EEG. Davis would tell Grass what kind of instrument was needed, sometimes in great technical detail; and Grass would figure out how to make it. As epilepsy centers proliferated throughout the country, the Grass Company flourished and became the leading commercial enterprise in manufacturing EEG machines.

Others in the Neurological Unit were formally honored. Frederick Gibbs and William G. Lennox won the 1951 Lasker Award for contributions to medicine, which many consider the American equivalent of the Nobel Prize. Lennox was called the "father of the fight against epilepsy."[10]

William G. Lennox

Lennox (1884–1960) had what might be considered the most romantic life of any of the team. In Colorado, his father was a gold miner, outfitter to prospectors, and cattle rancher.[11] A strike by one of his customers made him rich enough for other enterprises, including being president of a bank, and therefore rich enough for life. He was also a devout Methodist, divesting himself of some of his fortune by donating to schools and hospitals. Lennox himself aimed to become a missionary in China but failed because he had not studied Latin or Greek, requirements for admission to theology school. Instead, he devised another path to mission.

After graduating from Colorado College, he went to Harvard Medical School, graduated in 1913, and then interned at the Massachusetts General Hospital; in 1910, he had married Emma Buchtel, daughter of a former governor of Colorado. She encouraged him in his plans for a ministry among the poor of China, and in 1915 they moved to the Peking Union Medical College Hospital, where he stayed for four years. There, according to one obituary, the eight-year-old daughter of a friend had seizures, and he began his lifelong interest in the disorder. However, the two young Lennox daughters had repeated infections while in China, so they returned to Boston and Stanley Cobb in 1919. Lennox kept up his interest in China and returned there for a year of ministry in 1928.

Fred and Erna Gibbs: A Vascular Theory of Epilepsy

One of the first problems Lennox faced when he reported to Cobb was the need for a research assistant, so a want ad was published and answered by a young German woman, Erna Leonhardt. The results were profound. Erna was not only an assistant; she became an investigator herself. In 1929, they were joined by Fred Gibbs; he and Erna married the next year, and he became an integral part of the team with Lennox.

The research challenge was to determine what suddenly changed in the brain to cause a convulsive seizure, which often started abruptly. Lennox had devised a technique to take blood from the arteries that carry blood from the heart to the brain and from the internal jugular vein, which sends blood from the brain back to the heart. In sampling blood from these vessels and measuring the oxygen content in normal animals, he found that a measure called the "respiratory quotient" was 1. This particular ratio of carbon dioxide produced to oxygen consumed was determined to be that value for the oxidation of pure carbohydrate. For technical reasons, this result meant that the brain had to metabolize glucose itself, not fat or protein, individually or in combination. It also meant there had been no "anoxia" (with low levels of oxygen and high levels of carbon dioxide), which excluded one of the prominent theories of the day, that seizures were precipitated by a fall in oxygen content.

These early studies about glucose and the brain were later confirmed by Seymour Kety, Louis Sokoloff, and their colleagues at the University of Pennsylvania.[12]

There was still a possibility that seizures were caused by "cerebrovascular vasospasm," a sudden clampdown of the arteries that deliver blood, oxygen, and nutrients to the brain. To measure this, they needed a device that could record a sudden decrease in cerebral blood flow. After Fred and Erna married, they went to the University of Pennsylvania's Johnson Foundation for Medical Physics and developed the complicated needlelike instrument called a "thermocouple," which detects differences in heat and converts that information into measures of cerebral blood flow.

The Gibbs duo returned to Boston and set up the experiment in 1930. Lennox himself was the subject. Fred set up the blood flow detector as thin as a needle, and Erna inserted the sampling needle into a vein. Houston Merritt was there, too. He put an amyl nitrite capsule under Lennox's nose; this chemical had been used to treat angina of the heart by dilating blood vessels to increase blood flow. When Lennox inhaled, Gibbs recorded the expected increase in flow, but suddenly the record told them that flow had drastically declined. When they looked at Lennox, he was ashen pale and almost unconscious. He had inhaled the amyl nitrite a bit too enthusiastically and fainted. Fortunately, he recovered rapidly, and they considered Fred Gibbs's instrument a triumph. They subsequently did experiments that excluded vasospasm as a precipitant of seizures; in fact, blood flow was increased.[13,13a]

Gibbs[14] told an anecdote to emphasize the importance of this finding.

Penfield came down from Montreal to give an AOA lecture at Harvard. [AOA, or Alpha Omega Alpha, is an honor society, the medical school equivalent of Phi Beta Kappa.] Everybody turned out to hear Doctor Penfield. He was a great neurosurgeon. I knew him well, a really wonderful person. His talk was, of all things, on the role of cerebral vascular constriction in initiating seizures. Afterwards a lot of the students went to talk to him and I also came down to talk to him.

The AOA people were all around him and he called out. "Fred, what did you think of it?"

I said, "I thought you set these students' knowledge of epilepsy back at least 30 years."

The students were aghast. They wrote me a letter saying that I should write and apologize to Dr. Penfield. Before I could get that letter off, I received one from Penfield, apologizing to me, saying: "I didn't read those reprints you had sent me. It is all in there of course."[15,15a]

What was "all in there" included more than the evidence that seizures were attended by increased blood flow; the reports announced the fundamental usefulness of the EEG to the diagnosis and treatment of epilepsy.

Gibbs cited an experiment later done by Penfield in which he documented the increased cerebral blood flow during a seizure when he actually saw it with his own eyes during an operation on the brain of an epilepsy patient. Gibbs also noted that Penfield wrote the inscription for the Lasker Award given to Fred and Erna Gibbs with William Lennox: "Able and effective investigators in epilepsy whose devoted endeavors have saved a host of sufferers from misery."

An Early Classification of Epilepsy

Lennox continued to work on epilepsy. He classified seizures into three basic categories—grand mal, petit mal, and psychomotor—as defined by different clinical manifestations and different EEG patterns for each. He discovered that petit mal can be induced by overbreathing, which lowered the level of carbon dioxide in the blood; and he observed that this maneuver did not affect patients with grand mal or psychomotor attacks. He viewed all of them as chemical disorders, not psychological problems; and he found an effective drug for petit mal, trimethadione. He noted the genetic component in families with more than one person affected; this is now called a "complex pattern of inheritance" (to differentiate it from simple Mendelian forms of dominant or recessive inheritance).

Lennox the Humanitarian

Lennox was interested in epilepsy as a human tragedy on a personal level, but he also emphasized that it did not preclude a full and productive life. He was an advisor to the several voluntary agencies concerned with the condition. In 1944, he established the Seizure Unit at the Boston Children's Hospital, with four goals: diagnosis and treatment; training of specialists at all levels, including physicians, psychologists, social workers, nurses, technicians, and whatever else was needed; research; and public education to dispel myths about epilepsy and to foster the other goals. "Comprehensive epilepsy centers" are now functioning throughout the country, still following Lennox's model.

In 1960, hundreds of his admirers attended a dinner in honor of William Lennox.[15b] They presented him with the first copy of his new two-volume treatise on epilepsy.[16] He rose to give his own address of gratitude; but his voice faltered, and he had to turn the podium over to his daughter, Margaret, to finish his speech. His difficulty talking proved to be the first manifestation of a stroke that took his life, a heroic end of a heroic life.

Fred and Erna Gibbs Develop Electroencephalography

Fred and Erna Gibbs had worked closely with Lennox. The needle-like thermocouple probe they used to measure cerebral blood flow was actually constructed by Erna with her own skilled hands (to the admiring embarrassment of Fred). After Hallowell Davis had recorded the first EEG of normal brain rhythm in the United States (his own brain), the Gibbses were the first to record a petit mal seizure, and they induced Albert Grass to fabricate a practical EEG machine. In 1936, Fred and Albert drove to Kansas in a station wagon loaded with EEG equipment; this was the first U.S. exhibition of the new technique. Several German neurologists began to publish EEG records, and in 1935, Gibbs attended a meeting in Moscow. On the way home, he stopped in Germany to visit a neurologist named Hans Berger, who had been one of the first to publish detailed pictures of normal brain waves. Gibbs was surprised to find that Berger, being Jewish, had lost his university position and his equipment. He had continued to publish papers, but they were based on his notebooks and records, not new studies. In 1936 Erna and Fred translated Berger's abstracts into English. Hal Davis joined them in sending the index to Berger as a Christmas card. Gibbs later learned that Berger became seriously depressed and committed suicide by hanging in 1941.

One day, Dean Burwell of Harvard Medical School called Gibbs to his office and said that Gibbs was "the oldest instructor Harvard had ever had and that there were too many generals" in their department. Gibbs took that as an invitation to leave, heard about a position available at the University of Illinois, and moved. There, he and Erna maintained their positions as leaders of EEG and epilepsy, collaborating with neurosurgeon Percival Bailey in developing surgical treatment for what was originally called "psychomotor epilepsy" and became "temporal lobe epilepsy" to accommodate the change in concept about the origin of this class of seizures.

The departure of the Gibbs family was not, however, the end of the Neurological Unit; and the mantle would be taken first by Putnam and then by Merritt.

Notes

1. University of Pennsylvania School of Medicine. History, http://www.med. upenn.edu/history.html (accessed July 20, 2007).

2. Columbia University College of Physicians & Surgeons. History, http:// www.cumc.columbia.edu/dept/ps/descrip.html#history (accessed July 20, 2007).

3. Harvard Medical School. History, http://www.hms.harvard.edu/about/ history.html (accessed July 20, 2007).

4. Flexner A. I remember, the autobiography of Abraham Flexner. New York: Simon & Schuster, 1940, p 112.

5. White BV. Stanley Cobb: a builder of the modern neurosciences. Boston: Francis A. Countway Library of Medicine, 1984, Appendix A.

6. Ibid.

6a. Columbia University, History.

6b. Ibid.

7. Vilensky JA, Gilman S, Sinish PR. Denny-Brown, Boston City Hospital, and the history of American neurology. Perspect Biol Med 2004;47:505–518.

8. Stone JL, Hughes JR. The Gibbs Boston years: early developments in epilepsy research and electroencephalography at Harvard. Clin Electroencephalogr 1990;21:175–182.

9. Galambos R. Hallowell Davis (August 31, 1896-August 22, 1992). Biographical memoirs, vol 60. Washington DC: National Academies Press, 1998, pp 116–137.

10. Gibbs FA. William Gordon Lennox. 1884–1960. Epilepsia 1961;2:1–8.

11. Lombroso CT. William G. Lennox: a remembrance. Epilepsia 1988;28 (suppl 2):S5-S14.

12. Kety SS. The metamorphosis of a psychobiologist. Annu Rev Neurosci 1979;2:1–15.

13. Gibbs FA, Lennox WG, Gibbs EL. Cerebral blood flow preceding and accompanying epileptic seizures in man. Arch Neurol Psychiatry 1934;32:257–272.

13a. Galambos, Hallowell Davis.

14. Vilensky, Gilman, Sinish. Denny-Brown.

15. Gibbs FA, Davis H, Lennox WG. The electroencephalogram in epilepsy and in conditions of impaired consciousness. Arch Neurol Psychiatry 1935;34:1133–1148.

15a. Lombroso, William G. Lennox.

15b. Gibbs, William Gordon Lennox.

16. Lennox WG. Epilepsy and related disorders (2 vols). Boston: Little, Brown, 1960.

Tomorrow, I cross from New England and my life experience what is for me a Mason and Dixon line. Truthfully, I go with fear and trepidation.

Words attributed to Tracy J. Putnam at a Boston farewell party in honor of his appointment as director of the Neurological Institute of New York[1]

4

Putnam Moves from Harvard and Boston to Columbia and New York City: In Harm's Way

The Neurological Institute of New York was founded in 1909. Historical accounts laud it as the first hospital in the United States devoted to patients with neurological diseases.[2,3] But that was only partly true. Neurologists in those days were called "neuropsychiatrists," necessarily so. The specialty had originated during the Civil War, not even 50 years earlier. Neurological diseases were impervious to available treatments, and it is now difficult to think of a single condition other than epilepsy that could have been treated effectively with a drug or surgery. Another push for the hybrid term came from patients, who felt stigmatized by a psychiatric diagnosis and preferred to be admitted to a neurological hospital rather than an psychiatric hospital. Besides, psychiatrists were mostly in charge of asylums hidden in distant sites, remote from big cities. From the beginning and continuing for more than 50 years, many—perhaps even most—patients admitted to the institute probably had psychiatric disorders.

Since the 1950s, however, advances in neuroscience have totally transformed the specialty, moving treatment to center stage for many conditions, even though giant public health challenges remain to this day; Alzheimer disease claims more than 4 million people now, and Parkinson disease is closing in on another million.

The New York Neurological Institute is now part of a great medical center in uptown Manhattan, in a district called Washington Heights.

However, it was originally in a separate building in midtown Manhattan at 149 East 67th Street near Lexington Avenue. The founders of the institute were practicing neurologists named Joseph Collins, Joseph Fraenkel, and Pearce Bailey. It was a success by the standards of the time.

In 1915, Charles A. Elsberg was appointed chief of Neurosurgery. In 1922, Charles Wadsworth Schwartz became head of the new Department of Radiology.

Frederick Tilney (1876–1938) and Wilder Penfield (1891–1976)

Two of the most prominent figures in the early history of the institute were neurologist Frederick Tilney and neurosurgeon Wilder Penfield, who later became as prominent as Harvey Cushing, founder of brain surgery in the United States.

As a teenager, Tilney was already a cub reporter for an influential daily newspaper, the *New York Sun*, when he decided to become a doctor. He graduated from Yale with honors, worked his way through the Long Island College of Medicine by writing baseball stories for a boys' magazine, received his medical degree in 1903, was an intern at Kings County Hospital, and then practiced in Brooklyn. He started to do research and qualified for a PhD from Columbia in 1912; his thesis was on the cellular anatomy of the pituitary gland in different species. He became associate professor of neurology and neuroanatomy in 1914 and was a full professor the next year. He joined the staff of the Neurological Institute in 1920. Not an experimentalist, his research was observational, recorded in a monograph written with Henry Alsop Riley and called *Form and Function of the Central Nervous System*.[4] The monograph became a classic atlas of brain sections and was used by medical students for decades. Riley was regarded as a student of Tilney.

Tilney became a force in the Neurological Institute when it moved to the Columbia-Presbyterian Medical Center in 1929. He was credited with raising half the money needed to erect the new building and was chair of the Department of Neurology at Columbia University. From 1918 to 1921 he edited the *Neurological Bulletin of Columbia University*, which published case reports from the Vanderbilt Clinic, the ambulatory care facility of the neurologists and in a separate building before the move uptown. The journal was apparently not received enthusiastically, but Tilney tried again in 1931, when he edited a second journal, *The Bulletin of the Neurological Institute of New York*, which also had only a limited run. Nevertheless, Tilney's public eminence was documented in 1919, when he became physician to Adolph Ochs, owner of the *New York Times*. The medical problem would now be called "bipolar disorder."[5] Neurologists did practice psychiatry in those days.

Wilder Penfield made contact with Frederick Tilney while the Neurological Institute and Presbyterian Hospital were still separate buildings in Manhattan. Their relationship was not smooth. In 1921 Penfield returned from research training with two Nobel laureates: the British neurophysiologist Charles Sherrington and the Spanish cellular anatomist Ramon Cajal. Allen Whipple, chief of Surgery at Presbyterian, offered Penfield a position for cellular research, not surgery.

As Penfield told the story, a patient came in with a brain tumor, and Penfield called for a neurology consultant, who proved to be Tilney. Penfield recognized an attitude he had encountered in London, where the custom was simple and direct. The neurologist would tell the neurosurgeon when and where to operate. Penfield read the note Tilney had left in the patient's chart and called him. He thought Tilney was "rude" on the phone. The affair became a departmental issue, and Whipple asked Penfield to calm down because of impending negotiations. Tilney wanted all neurology and neurosurgery to be carried out in the Neurological Institute; Whipple wanted some activity in Presbyterian Hospital. The two buildings were separate, with the hospital at Park Avenue and 70th Street and the Neurological Institute in a rented building at 57th and Lexington. Penfield had to cross Central Park from the hospital on the East side to teach surgical pathology at the College of Physicians and Surgeons, the medical school on the West side. Tilney had joined the institute staff in 1920 and played a dominant role in subsequent developments.

Penfield felt as though he were "a pawn on Tilney's larger chessboard" and recognized Tilney's power. "When a pawn is in the way, it can be moved or sacrificed. He elected to move me."[6] Penfield therefore joined the institute staff in 1922 and would report to Charles Elsberg, the first head of neurosurgery. He had the privilege of transferring patients from Presbyterian Hospital. Penfield thought,

Charles Elsberg kept a jealous eye on younger men . . . he had little interest, and indeed little understanding, of the science that should, it seemed to me, be basic to neurosurgery. . . . The other assistant surgeon working under Elsberg in 1922 was Byron Stookey. He was not happy in this relationship. He himself was an impressive man, a little older than I was, trained at Harvard and Paris . . . At first, [Henry A.] Riley impressed me and gave me a feeling of inferiority, as Stookey had. But, in time, these two men were to become my close friends and delightful companions at meetings in many parts of the world.[7]

Penfield wearied of working at both the hospital and the Neurological Institute. He was a pathologist but had no microscope at the institute, so he had to return to Presbyterian Hospital to use one. He asked Whipple if he could work only at the hospital, but he wrote, "The real situation was that I could not accept dictation from Elsberg, and that the chief neurologist, Tilney, did not want to use me. It was agreed that I would confine my work to PH."[7a] He understood that, after the move to 168th Street, his patients would be transferred to the new institute.

He said that, after that phone conversation about a patient, Tilney was never rude to him again, but their relationship was cold. He thought he had more support from Stanley Cobb in Boston than from Tilney and Elsberg combined. He said:

Elsberg "*might* have a place for me." That was his attitude. It was quite understandable that Elsberg should wonder vaguely what to do with me when the time should come for the NINY to take over neurosurgery at the new medical center. During the transition period of the past six years, 1921–1927, I had served a transitory purpose to the Presbyterian. Even at the institute, while I had taken no clinical responsibility, I had been their neuropathologist for almost three years, and this was, of course, without pay.

I liked Elsberg for his enthusiastic interest in each clinical problem. But that was all we had in common. Somehow, I felt shy with him. I was like a horse left free too long to gallop in the pasture.

So it was that, as the year of 1928 was ushered in, I was laying plans to move in three months time, with my surgical and medical associates, from the old PH to the new medical center uptown at 168th Street and Broadway. The NI was scheduled to follow, moving into their own neighboring, newly built institute a year later, March 1929. Until that time, the freedom I had enjoyed for the past six years at the Presbyterian would continue to be mine, freedom to study patients and to treat them, freedom to make the new approaches to the unsolved problems of neurology that occurred to me.

After that date, it was intended by those who had overall control of the hospital and medical school that [William] Cone and I would shift our allegiance to the institute. But, since no one in the NI who could speak with authority (least of all Professor Tilney) had taken me into his confidence or suggested that he needed me, my departure from New York was inevitable.[7b]

That is how Columbia University, Presbyterian Hospital, and the New York Neurological Institute lost the neurosurgeon who founded the Montreal Neurological Institute, who started surgery for the treatment of epilepsy, who established the concept of temporal lobe epilepsy, and who is in the pantheon of great neurosurgeons alongside Harvey Cushing.

Edward S. Harkness (1874–1940) and the Creation of the Columbia-Presbyterian Medical Center

Intrepid: resolute fearlessness, fortitude and endurance

Despite his problems with Penfield, Tilney pressed on. He was a nationally important figure; he was one of six founding editors of the first American journal for neurology, the *Archives of Neurology and Psychiatry*.[8] It is said that he personally raised more than half of the $2 million needed for construction of the new Neurological Institute building on 168th Street.[9] The architect was James Gamble Rogers, who also designed the Yale University campus in New Haven. The genius behind the institutional mergers was Edward S. Harkness, the Rockefeller lawyer who prevailed over the durable but unreliable and obstructionist president of Columbia University, Nicholas Murray Butler (1862–1947).[10,11] Butler was a Nobel Prize winner for Peace and president of Columbia University from 1902 to 1945. He was an effusive orator at the 1925 groundbreaking ceremonies for the medical center, stating that "we propose to build a monument more lasting than bronze which shall testify alike to the growing power of human knowledge to minister to the physical and mental ills of man and to the zeal of civilized man to help and to cure his less fortunate fellows."[12]

However, in 1915, Butler reneged on promises to raise half the costs for the Highlander baseball property in Washington Heights. In 1917 he refused to join Presbyterian in establishing the medical center, and in 1918 he declared that Columbia lacked funds. In essence, Butler repeatedly failed to honor financial commitments and rejected plans for the location of the hospital. In fact, negotiations between the university and hospital did not proceed until Butler was prohibited from attending the meetings.[13]

Edward Harkness came to the rescue. He was the son of Stephen S. Harkness, who started as a harnessmaker but ultimately became a partner of John D. Rockefeller and the second largest shareholder in the Standard Oil Company. Edward, the son, became one of the 10 wealthiest men in the history of the United States and was a philanthropist of the highest order; buildings funded by him or his mother and named for them grace the campuses of Yale, Harvard, Brown, Columbia, Oberlin, Connecticut College, Phillips Exeter Academy, St. Paul's School, and Lawrenceville Academy. James Gamble Rogers was chosen to be the architect for all of these projects. Edward's mother also established the Commonwealth Fund, "to do something for the welfare of mankind";[14] and her son was its first president. The fund is still active in medical education, research, and health-care policy but has also been involved in child guidance, mental health, rural hospitals, and another of Edward's favorite themes, fellowships to strengthen links between America and England.[15]

The site for the new medical center had been the home of a baseball team born in controversy. Competition between the National and American Leagues started early. The New York Giants of the National League were ensconced in the Polo Grounds, near 155th Street in Manhattan; and they objected to plans for a second New York team. Instead, the organizers of the American league created the Baltimore Orioles, which began playing in 1901. The next year the Giants began to raid the Orioles for the best players, and the feud grew ugly. In a gesture of peace, the Baltimore team was allowed to move to New York. In 1903, their home became Hilltop Park, which was rapidly but shoddily constructed for them between Broadway and Fort Washington Avenue from 165th to 168th Streets.

The 22 acres were ample, sited high on a plateau above the neighboring Riverside Drive and the Hudson River. Accordingly, they changed the team name to the Highlanders. The next decade was marked by ups and downs. In 1911, the Polo Grounds burned down and the Highlanders allowed the Giants to play in Hilltop, an act of kindness that paid off in 1913 when the uptowners were allowed to play in a newly renovated Polo Grounds at 155th Street. No longer at a lofty altitude, the Highlanders became the New York Yankees in 1913, and the team moved from the Polo Grounds to Yankee Stadium in the nearby Bronx in 1923. The Hilltop stadium was demolished in 1914, and evangelist Billy Sunday built a tabernacle there; it lasted for a decade and was replaced by the medical center.

In 1929, Presbyterian Hospital and the Neurological Institute moved to Washington Heights together. There, they were joined by Babies Hospital, the Squire Urological Hospital, the Sloan Hospital for Women, and the Columbia College of Physicians and Surgeons. This was the first joint project in the United States of a medical school and its affiliated hospitals to create an "academic medical center"; the Columbia-Presbyterian Medical Center became the model for other medical schools. Financial integration of the separate hospitals with Presbyterian Hospital was achieved gradually and was not completed until the New York Orthopedic Hospital merged in 1950.

Soon after the 1929 move, however, the Neurological Institute was faced by "yearly deficits [that] had steadily increased."[16] Controversies arose about Presbyterian's charges to the institute for utilities. In October 1935, the institute was in arrears almost $42,000 and stopped payments for some time because, they claimed, Presbyterian was overcharging for steam. In December 1936, however, the managers of Presbyterian approved a settlement of $25,000 on the outstanding charges of some $37,000. As a result, in 1937, the Presbyterian managers took control of the institute, appointing Walter Palmer as professional director of the institute in addition to his position as medical director of Presbyterian Hospital. That was the end of the independent institute, but it gained in many ways by becoming a unit of a general hospital. For instance, the amalgamation facilitated consultations for neurology patients by specialists in internal medicine, general surgery, or surgical specialties.

Frederick Tilney Becomes Director of the New York Neurological Institute

From the time of the merger, Frederick Tilney played a conspicuous role. However, in 1924, when he was only 48 years old, Tilney had an ischemic stroke in the left hemisphere of the brain, resulting in weakness of the right arm and leg and "aphasia," difficulty speaking. His recovery was slow, and it is not clear how complete it was. Indomitable as he was, however, he learned to write with his left hand and he limped thereafter. Despite these handicaps, he was able to keep on working and writing by hand so that 1928 saw the publication of his two-volume anatomic treatise *The Brain from Ape to Man*. In 1926, he was president of the American Neurological Association, another measure of the high esteem his peers bestowed. In 1931, he founded a short-lived journal, the *Bulletin of the New York Neurological Institute*. After he died, obituaries lauded his neurological publications, his role at the Neurological Institute, and his observations of criminals.[17] *Time* magazine claimed that he had had "an influence among devoted young neurologists second only to that of famed" Harvey Cushing.[18]

Formally, Tilney became medical director of the Neurological Institute in 1935. However, he died of heart disease in 1938, and the intervening three years must have been medically difficult for him—and for the institute. An associate director, C. Burns Craig, had been appointed simultaneously, presumably to mollify Tilney's handicaps; and Tilney resigned "early in 1938."[19] Despite the permanent effects of the 1924 stroke, he had continued his national activities; but the incapacitating heart disease caused him to resign from the editorial board of the *Archives of Neurology and Psychiatry* in 1935.[20] That was also the year he resigned as director of the Neurology Service, transferring responsibility to Walter Palmer and then to Robert Loeb, as associate medical director; they were specialists in internal medicine and not neurologists, a clear sign of serious problems in the Department of Neurology and the institute. That state of affairs lasted until Putnam arrived in 1939. Tilney had served as director from 1920 to 1935, followed by Palmer from 1935 to 1939;[20] Palmer was already fully occupied by being the chair of Medicine, and he relegated neurological responsibilities to another professor of Medicine, Robert Loeb, who served until Putnam arrived.

Meanwhile, in the Department of Neurosurgery, Byron Stookey was the head from Elsberg's retirement in 1937 until 1939, also awaiting the arrival of Putnam. Elsberg had been chair for 22 years, and he may have been ready for retirement; but it is said that Stookey did some pushing to gain the chair himself.[21]

Encephalitis Lethargica: A Failure of Medical Research

Tilney published little about human diseases, but he became heavily involved in the epidemic of "encephalitis lethargica," or sleeping sickness, which accompanied the worldwide pandemic of influenza in 1918 and was blamed for postencephalitic parkinsonism. The cause of the acute infection was never discovered,[22] not even after later attempts with modern viral DNA studies[23] and molecular antibody studies.[24] If even advanced technology failed, we cannot criticize the failure of Tilney and his associates for the lack of progress. Indeed, a dominant theory was that the brain inflammation was part of a whole-body reaction to "focal infection." Many patients were "treated" by pulling out all their teeth to eliminate the supposed infection.

The first meeting of the Association for Research in Nervous and Mental Disease, held in 1920, produced a monograph on encephalitis. At the Rockefeller Institute, Simon Flexner set out to find the virus; he never succeeded. Columbia and Tilney were drawn in by William J. Matheson, who had made a fortune in manufacturing chemicals, contracted influenza, and ended with a diagnosis of parkinsonism. He approached William Darrach, dean of the College of Physicians and Surgeons, and donated $10,000 annually for encephalitis research. Tilney was one of Matheson's physicians, and he enrolled Haven Emerson, director of the New York City Health Laboratories. Josephine B. Neal was Emerson's assistant, and she was given the task.

One historian[24a] analyzed Neal's challenge:

Her position as a woman made her somewhat of an outsider within the overwhelmingly male circles of medical administration, practice and research—yet she emerged as the single most important member of the Commission. Her willingness to conduct the massive amount of bibliographic spadework that Dana and Tilney had previously deemed necessary for such a project eventually brought her out of the [New York Academy of Medicine] library to become the most knowledgeable clinical expert on encephalitis in the wards of the Neurological Institute. [She had an] unusual combination of talents—she was a neurologist with expertise in bacteriology rather than neuroanatomy.[24b]

The focal infection theory was championed at the Mayo Clinic by Edward C. Rosenow. Matheson went to the famous clinic and lost his teeth without benefit. Rosenow journeyed to Columbia a few months later in 1928 to study body fluids from patients, seeking streptococci to culture. Another celebrity to join the fray was J. P. Morgan, whose wife died of encephalitis; he donated $200,000 to build a floor in the Neurological Institute. Meanwhile, Matheson was given streptococcal vaccine and reported that he felt much better. (There were no placebo-controlled therapeutic trials in those days, not for any condition.) Neal and other experts publicly attacked Rosenow, offending Matheson in a contretemps that took effort to repair. Tilney was the leader of all the research and kept sending samples to the virologists, though he spared enough money to assuage Rosenow, too.

However, when it was found that streptococci could be isolated from asymptomatic people, members of the commission decided in 1930 to end support for Rosenow's activities in New York. That year, Matheson died, leaving a large fund to endow the commission. Willard C. Rappleye succeeded Darrach as dean. In 1938, a sick Tilney resigned from the commission. Support for Neal's clinical studies ceased in favor of laboratory virology. Her third report, in 1939, generated no new source of money. That year, Putnam arrived, expressed his skepticism about the research, and noted that the syndrome had mostly disappeared. In the meantime, 1000 patients had been treated with the Rosenow streptococcal vaccine or a herpes simplex vaccine.[25] No one was helped, but at least no major adverse effects resulted.

The Interregnum

After the encephalitis fiasco, serious problems arose. Tilney had permanent effects of the 1924 stroke, and he was later incapacitated by heart disease, which caused him to resign from the editorial board of the *Archives of Neurology and Psychiatry* in 1935. That same year he also resigned as director of the Neurology Service, transferring responsibility to Walter Palmer and then to Robert Loeb. These internists took over direction of the Institute and Stookey was pushing Elsberg. That lasted until Putnam arrived in 1939. Meanwhile, in the Department of Neurosurgery, Byron Stookey was chief from the time of Elsberg's retirement in 1937 until 1939, also awaiting the arrival of Putnam. Elsberg had been chair for 22 years and may have been ready for retirement, but it was said that Stookey did some pushing to gain the chair himself.[26]

The Birth of Neuroradiology: Dyke and Davidoff

The year of the move, 1929, was notable for another reason: The Neurological Institute was soon to provide national leadership in the development of medical specialties related to neurology and neurosurgery, notably neuroradiology and neuropathology. In that year, Cornelius G. Dyke became the first full-time radiologist for the institute. After a one-year internship, he spent one year at the Peter Bent Brigham Hospital of Harvard with Merrill Sosman, a general radiologist who worked with Harvey Cushing; and in so doing, he became a founder of neuroradiology. Dyke continued his most productive years in neuroradiology at Columbia for 13 years. In 1942, he was diagnosed with acute myelogenous leukemia, and he died in April 1943, at age 43; that disease may have been caused by occupational exposure to radiation.

Dyke had teamed with neurosurgeon Leo M. Davidoff, another Cushing trainee, to write monographs on normal and pathological findings seen on X-rays of the skull. Like Penfield before them, they used "air studies" but, unlike Penfield's early forays, they became experts in the procedures that provided the first photographic X-ray images of the brain itself. They injected air into the cerebrospinal fluid (CSF) by lumbar puncture at the base of the spine or by inserting a needle through the skull and brain into the ventricles, the large pools of fluid deep within the brain. The appearance of air and brain tissue differs in X-ray films, and distortions of the normal pattern could identify brain tumors or abnormalities of the CSF circulation. The method was a forerunner of computerized tomography (CT) and magnetic resonance imaging (MRI) but with less detail, more pain for the subject, and more risk of serious harm to the patient. It was an important advance for the time, but it was a boon for medicine, science, and humanity when, some 40 years later, CT and MRI replaced the injection of air.

Charles Elsberg (1872–1948): Another First-Generation Neurosurgeon

Elsberg himself was a unique personality. He attended public schools in Manhattan, including the scholastically elite Townsend Harris High School, and then City College. He was a graduate of Columbia's College of Physicians and Surgeons and became a general surgeon, gradually restricting his activities to neurosurgery, just as the other first-generation neurosurgeons did—including Cushing himself, Charles Frazier in Philadelphia, and Earnest Sachs in St. Louis. They worked together to found the Society of Neurological Surgeons, and Elsberg later became president of that club. His first surgical appointment was at Mt. Sinai Hospital in New York. In 1913, he published a review of spinal surgery, and his first book on the subject was published in 1916. His second book on spinal cord surgery, published in 1925, included one of the first descriptions of the neurological consequences of a "slipped disc" and the benefits of surgical treatment. A third volume on spinal cord surgery, published in 1941, included chapters on radiology by Cornelius Dyke and pathology by Abner Wolf and was considered an "indispensable reference book for a generation of neurosurgeons."[27]

Elsberg was recognized by neurologists as well as neurosurgeons when he was made president of the American Neurological Association. Among his other kudos, he received an honorary degree from City College in 1947, together with Supreme Court Justice Felix Frankfurter and Senator Robert F. Wagner.

All of those pioneer neurosurgeons had powerful personalities. Elsberg seems to have been no exception.[28] Lawrence Pool, who became chief of Neurosurgery at the Neurological Institute in 1949, had earlier been a resident with Elsberg. Pool wrote that Leo Davidoff was "Elsberg's favorite first assistant in the operating room."[29] Nevertheless, Paul Bucy, a notable Chicago neurosurgeon, edited a book containing a biosketch of Elsberg and quoted Davidoff:

"I can say nothing good about the bastard and I won't say anything bad."[30]

Bucy also added that Byron Stookey spoke to him for hours about what a despicable person Elsberg was. Pool also wrote that Stookey proclaimed "in tones well above a whisper" that "He can't make or confirm a diagnosis until he looks in the patient's chart!" Pool continued: "Dr. Stookey even took steps to prevent Dr. Elsberg from having access to the charts. This bit of pettiness frankly did not sit well with those of us who were interns."[30a]

Nevertheless, Elsberg established a prestigious fellowship to honor his wife, and the New York Neurosurgical Society set up an annual lecture in his name. Stookey replaced him as head of neurosurgery at the institute in 1937 but had that position for only two years. Elsberg died in 1948.

Byron Stookey (1887–1966):
Domineering Neurosurgeon

If, as it seems, Stookey was determined to end Elsberg's career, he was also nefarious in Putnam's misadventure at the institute. An imposing six foot four inch athlete, he was, by all accounts, a fearful figure to anyone in a subordinate position. He graduated from the University of Southern California in 1908 and gained a master's degree from Harvard in 1909 and an MD there in 1913. He was an intern in general surgery at the Boston City Hospital and then served in the British army during World War I. In 1919 he went to Bellevue Hospital in New York, concentrating on neurosurgery. Pool said that Stookey was one of the last well-known American neurosurgeons to have had no formal preliminary training in the specialty. He joined the staff of the Neurological Institute in 1922, when it was still on East 67th Street, where he first assisted Elsberg. He became a highly respected surgeon. Pool soon made a social error in dealing with Stookey; he was late in returning from a vacation. Stookey therefore rejected Pool's application for staff privileges. Stookey, Pool wrote,

> seemed to attract controversy like a lightning rod that attracts lightning on a clear day. Perhaps this was because, despite being an extremely capable diagnostician and top-of-the-heap surgeon, he tended to let colleagues know it. Nor did his dictatorial manner help.
>
> His overbearing attitude finally led to trouble. The Institute's neurologists and other non-surgical members of its staff, protested to such an extent that in late 1939, the governing Board of the Medical Center asked him to step down as chief of Neurosurgery. Politically the non-surgical staff was then in turmoil because it had no leader [considering Tilney's stroke and heart disease].[30b]

Boston City Hospital: Cradle of American Neurology

The impact of the early Neurological Institute has to be placed in the perspective of other centers in the 1920s and 1930s. Modern neurology emerged from three centers in the early years of the twentieth century: Philadelphia, New York, and Boston. By 1938, Boston seemed to have outpaced the others in neurological research. In New York, Tilney had published on comparative neuroanatomy and, with Henry Riley, produced a widely used and highly respected brain atlas. However, patient-oriented or disease-focused research was invisible at the Neurological Institute, except for the work of Elsberg, Davidoff, and Dyke. They formed a group that led in spinal cord surgery and the beginnings of diagnostic neuroradiology. But the Neurology Department was less active.

In Boston, the contrast was impressive. The neurology leader was Stanley Cobb, and all research in the department radiated from his interest in epilepsy, which took him to studies of cerebral blood flow and metabolism. Self-confidently, he gave his faculty members freedom to follow their own interests and publish; he managed to keep together a team that included Putnam and Merritt; Fred and Erna Gibbs, who developed the electroencephalogram; William Lennox, the dean of epilepsy research; Frank Fremont-Smith, a pioneer in characterizing the constituents of CSF; and Raymond Morrison, an early neuropathologist.

Boston City Hospital was the premier training program among the few centers actively turning out new neurologists. Merritt himself arrived in 1928 and stayed on. Among the others were Frank Forster (who chaired departments at Georgetown and Wisconsin), Charles Aring (who led the University of California, San Francisco, briefly and Cincinnati for decades), Harold Wolff (Cornell), Augustus Rose (UCLA), Joseph Foley (Metropolitan Hospital, Cleveland), Milton Rosenbaum (Psychiatry at Albert Einstein College of Medicine), Robert Aird (University of California, San Francisco), and Adolph Sahs (University of Iowa).

In New York, however, the Neurological Institute could not claim to be a national training center in the 1930s and had to await the arrival of Houston Merritt to warrant that role.

Meanwhile, Cobb gradually but steadily became more interested in psychoanalysis than he was in neurology. He had set the pattern in neurology to parallel that of academic internal medicine: Clinical research must be the core of a leading department because new knowledge includes the best possible treatment of patients. In 1934, Cobb moved to become head of Psychiatry at the Massachusetts General Hospital. To replace him, Tracy Putnam was appointed chief of Neurology at the Boston City Hospital.

Here, we come to a riddle. Why, five years later, did Putnam so eagerly accept the invitation to move to New York when he was already leading the most applauded neurology center in the country? Later, in New York, he was to be hounded by a powerful and aggressive neurosurgeon. Could that have been also true in Boston? Surely, Donald Munro was an aggressive neurosurgeon, but the records give no signs of overt conflict between the two leaders of neurology and neurosurgery at the Boston City Hospital.

Nevertheless, some comments in obituaries for Donald Munro (1889–1973) could have been written about Stookey. "W. W." (Walter Wegener) wrote one in which he described Munro's career from training in general surgery with Alton Ochsner in New Orleans and then neurosurgery with Charles Frazier at the University of Pennsylvania.[31] After army service during World War I, he joined the staff of Boston City Hospital in 1930, starting the neurosurgical service there together with Putnam under the aegis of a grant from the Rockefeller Foundation. He focused first on head injuries. He started a neurosurgical training program.

> He was good to his house officers while also being very demanding.... Two weaknesses were never tolerated: sloth and dishonesty. Sometimes he was regarded as a hard taskmaster, but when the training was over, he was loved the more for having been so.[31a]

Sir Ludwig Guttmann, in England, had established a model center for the treatment of spinal cord injury; he praised Munro's contributions to the field, especially his attention to infection and bladder function. He added, "In his pioneer work he had often to swim against the stream of prejudice and old concepts, which made him sometimes unpopular among his colleagues because of his forthright criticism." Munro railed against the "exploitation" of the public "in the name of teaching, the training of young doctors, and the colossal expenditure for investigation at the expense of patient care." He hoped that "the teaching of medicine will soon crawl out of the mysterious and gloomy wood of 'research' and return to the proper business of doctors—the care of the sick."[32]

His partner, Tracy Putnam, was the epitome of a research academician. Did they really get along? Was Putnam pushed from Boston as much as he was attracted to New York?

At a farewell dinner for Putnam, Munro said the honoree was making a mistake and warned Putnam with the following barnyard story:

[A farmer] called his veterinarian one night because the cat was sick. The veterinarian advised castor oil, which the farmer dutifully administered to the cat. The next morning the veterinarian met the farmer and asked how the cat was doing. The farmer said that he had trouble getting two teaspoonfuls down the cat's throat. The veterinarian replied: "Two teaspoonfuls! I said two drops. Is the cat all right?" The farmer said that she had diarrhea all night, but she had friends. "Twelve neighborhood cats showed up. Six of them were digging holes and the other six were covering them up."[33]

Munro went on to say, "Tracy, here you have friends to dig and cover up for you; you won't have friends like that in New York."[33a]

Notes

1. Schlesinger EB. Neuro chiaroscuro. "All in the Family" about Byron P. Stookey (1887–1966). Bull Alumni Presbyterian Hospital, 1996.
2. Elsberg CA. The story of a hospital, the Neurological Institute of New York, 1909–1938. New York: Hoeber, 1944.
3. Riley HA. The Neurological Institute of New York; the first hospital in the western hemisphere for the treatment of disorders of the nervous system. Bull N Y Acad Med 1966;42:654–678.
4. Tilney F, Rilery HA. Form and function of the central nervous system. New York: Hoeber, 1920.
5. Tiffit SE, Jones AS. The trust: the private and powerful family behind the New York Times. New York: New York Times Company, 1999, p 124.
6. Penfield W. No man alone: a neurosurgeon's life. Boston: Little, Brown, 1977, p 79.
7. Penfield, No man alone, 80.
7a. Penfield, No man alone.
7b. Penfield, No man alone, 82.
8. Casamajor L. Archives of Neurology and Psychiatry. In Fishbein M, ed. A history of the American Medical Association 1847–1947. Philadelphia: Saunders, 1947, pp 1149–1153.
9. Mettler FA. Frederick Tilney (1876–1938). In American Neurological Association (1875–1975). New York: Springer, 1975, pp 167–173.
10. Rowland LP, Ginsberg D, Abramson M, Erlanger BF, Turino GM, Yudofsky SC. P & S: an historical perspective. In Gerst SR, ed. The student handbook of the College of Physicians and Surgeons of Columbia University 1982-83. New York: P & S Club of the College of Physicians and Surgeons, Columbia University, 1982, pp 171-181.
11. Lamb AR. The Presbyterian Hospital and the Columbia-Presbyterian Medical Center, 1868–1943. New York: Columbia University Press, 1955, pp 126–136.

12. Wooster JW Jr. Edward Stephen Harkness 1874–1940. New York: Commonwealth Fund, 1949, p 62.
13. Rosenthal M. Nicholas miraculous. The amazing career of the redoubtable Dr. Nicholas Murray Butler. New York: Farrar, Straus and Giroux, 2006, p 242.
14. Wooster, Edward Stephen Harkness, 43.
15. Rosenthal, Nicholas miraculous, 51.
16. Elsberg, The story of a hospital, 133.
17. Frederick Tilney, noted neurologist, dies. He did much research work in insanity—made special study of criminals [obituary]. New York Times, August 8, 1938.
18. Anonymous. Medicine. Bread-&-butter brains. Time, October 16, 1939, http://www.time.com/time/searchresults/
19. Elsberg, The story of a hospital, 125.
20. Penfield, No man alone.
21. Kaufman HH, Goodrich JT. Byron Stookey: "the old lion" an unsung giant of neurosurgery. Neurosurgery 1997;40(2):383–388.
22. Kroker K. Epidemic encephalitis and American neurology, 1919–1940. Bull Hist Med 2004;78:108–147.
23. Reid AH, McCall S, Henry JM, Taubenberger JK. Experimenting on the past: the enigma of von Economo's encephalitis lethargica. J Neuropathol Exp Neurol 2001;60(7):663–670.
24. Dale RC, Church AJ, Surtees RA, Lees AJ, Adcock JE, Harding B, Neville BG, Giovannoni G. Encephalitis lethargica syndrome: 20 new cases and evidence of basal ganglia autoimmunity. Brain 2004;127(Pt 1):21–33.
24a. Kroker K. Epidemic encephalitis.
24b. Rosenthal, Nicholas miraculous, 51.
25. Louis ED. Vaccines to treat encephalitis lethargica. Human experiments at the Neurological Institute of New York, 1929–1940. Arch Neurol 2002;59(9):1486–1490.
26. Kaufman, Goodrich, Byron Stookey.
27. Horwitz NH. Charles A. Elsberg (1871–1948). Neurosurg Online 1997; 40(6):1315–1319.
28. Cramer F. Charles A. Elsberg. In Bucy P, ed. Neurosurgical giants: feet of clay and iron. New York: Elsevier, 1985, pp 355–360.
29. Pool JL. Brain anecdotes, 1932–1972. Torrington, CT: Rainbow Press, 1994, p 55.
30. Cramer F. Charles A. Elsberg. In Bucy P, ed. Neurosurgical giants: feet of clay and iron. New York: Elsevier, 1985.
30a. Kroker, Epidemic encephalitis.
30b. Pool, Brain anecdotes.
31. Signed "W. W." (Wegener W). Donald Munro M.D. 1889–1973. N Engl J Med 1973;88(19):1022–1023.
31a. Ibid.
32. Guttman L. Obituary, Dr. Donald Munro. Paraplegia 1973;11(2);197–198.
33. White BV. Stanley Cobb: a builder of the modern neurosciences. Boston: Francis A. Countway Library of Medicine, 1984, p 407.
33a. Ibid.

Hubris: overweening pride, self-confidence, superciliousness, or arrogance, often resulting in fatal retribution.

Wikipedia, http://en.Wikipedia.org/wiki/Hubris[1]

The Greeks had a word for it: hubris, the self-confidence and arrogance that always leads to disastrous retribution.

Sydney Brenner, 2007[2]

5

Putnam at the New York Neurological Institute: Clinical Activity and Vulnerability

There was never anyone as versatile neurologically as Tracy Putnam, not before or since. When he arrived in New York, he had estimable credentials in both neurology and neurosurgery, even also in psychiatry.

The closest comparison might have been the great Wilder Penfield, who called himself a "surgical neurologist." However, Penfield largely restricted his operations to studies of—and treatment for—epilepsy. And he created a great neurological institute in Montreal, where he encouraged Herbert Jasper to join him as electroencephalographer. Together they essentially invented the concept of "temporal lobe epilepsy," in which the seizures include abnormalities of thought and behavior. These attacks start in areas of the brain that were thought to have been damaged by birth injuries or other trauma. By ingenious surgery, he could excise the seizure focus and relieve the attacks, even though adding more traumatic damage in the process of operating. More than any historical figure in medicine, Penfield showed that the mind is in and of the brain. By stimulating different parts of the brain with an electrical current, he could evoke visions, sounds, and spoken words. Yet, he ended his life as a dualist, believing in a soul separable from the brain. But Penfield did not take on all of neurosurgery himself. Instead, he enlisted other excellent neurosurgeons to deal with brain tumors, hemorrhage, and traumatic brain injury.

As a neurosurgeon, Putnam had solid credentials. He had been trained by Harvey Cushing. He had been one of the three founders and the fifth president of the Society for Neurological Surgery (The Cushing Society). He had written papers on the treatment of brain trauma and disorders of the circulation of brain fluids ("hydrocephalus"), and he had devised operations to relieve tremors and other involuntary movement disorders.

As a neurologist, Putnam also had credentials. He was one of the founders of the American Board of Psychiatry and Neurology, the organization that sets training requirements and examinations for physicians entering these specialties. When he came to New York in 1939, he was editor of the *Archives of Neurology and Psychiatry*, the leading American journal for neurologists; Putnam held that position from 1935 to 1955, which made him a highly regarded and influential voice for neurology. By then, he had also started his research in multiple sclerosis, which was not a surgical problem.

As a psychiatrist, Putnam was the first president of the American Society for Psychosomatic Medicine in 1943. He already had personal experience in psychiatry. In one of his own biosketches he wrote, "In 1925, I went to Amsterdam to study neuropathology, and was assigned an anatomical problem which involved the presentation of information in three dimensions. I also studied psychology and psychiatry there."[3]

Putnam's first wife was Irmarita Putnam.[4] In 1917 and 1918, she was a friend of the daughter of Tracy's uncle, James Jackson Putnam; and she lived in their home. She could have met Tracy then or later, when he was an intern in surgery at Johns Hopkins and she was a medical student there. She became a psychiatrist and, in 1923, went to Amsterdam for a training analysis with J. H. van der Hoop. Tracy accompanied her and joined her as a pair of subjects in psychoanalysis, with the same therapist for about six months. Later, Irmarita was in analysis with both Jung and Freud. When she went to Vienna, she took her three-year-old daughter with her.

Sadly, in 1947, when their daughter was 20 years old, according to a news item in the *New York Times*, the daughter "fell or jumped" to her death when she was a student at Vassar College.[5] The report mentioned that the Putnams had already been divorced and that he had married again.

When Tracy Putnam came to the Neurological Institute in 1939, the hospital was in disarray. Its leader, Frederick Tilney, had been ill and inactive. The person in charge was Robert Loeb, brilliant for sure but not a neurologist. The only research of note was being done in neurosurgery and neuroradiology by Leo Davidoff and Cornelius Dyke, who set out to define the limits of normal variations in skull X-rays so that diagnostic abnormalities could be identified. The paucity of neurological disease-oriented research contrasted with the Neurological Unit at the Boston City Hospital, where Putnam, Merritt, Lennox, Fremont-Smith, and Fred and Erna Gibbs together invigorated epilepsy research. And Merritt had worked with others to write books that became classics on neurosyphilis, a scourge of the times, and on disorders of the cerebrospinal fluid. The institutional contrast was stark. Putnam later wrote that the Neurological Institute at the time was the "laughing-stock of the profession."[6]

Putnam was attracted to New York because the neurological service was "six times larger" than the one in Boston. He presumably also had a larger budget and he could hire staff. However, he brought with him from Boston only one colleague, Paul F. A. Hoefer, a German émigré, who set up the first electroencephalography laboratory in New York City. Robert Aird, who later became chair of Neurology at the University of California in San Francisco, had known Putnam in Boston. He thought Putnam should have brought Merritt as an assistant director,[7] but Merritt was the leading local candidate to succeed Putnam as head of the service at the Boston City Hospital. There is no evidence that Putnam ever considered this plan, and Merritt did not come to New York until 1945.

Putnam must have been busy from the day he arrived. He summarized the situation as follows:

[T]he war in Europe broke out just as I arrived. I was appointed to the National Research Council, to the Health Insurance Plan, to the Board of Neurology and Psychiatry and that of Neurosurgery, and became the neurological consultant to a group from Columbia investigating the effects of altitude and anoxia in one of the early pressure changers. All of this cut down my opportunities for doing personal research work.[8,8a]

The military draft for World War II must also have limited the number and quality of applicants for residency training positions.

Clinical practice seems to have evoked little discussion and was presumably satisfactory, perhaps taken over by Henry Riley and Edward Zabriskie. Putnam himself was recognized by the public. He was called in 1945 when the adolescent son of John Gunther was found to have signs of a brain tumor. The 16-year-old boy was away from home at a boarding school in Deerfield, Massachusetts. Gunther was a prize-winning author of the blockbuster best-selling political books *Inside U.S.A.* and *Inside Germany.* In April 1946, he was midway through the manuscript of *Inside U.S.A.* when he and his wife were called by the school physician, urging them to contact Putnam for signs that seemed to be those of a brain tumor. They found Putnam, and he rode with them by car for several hours to examine the boy at school. As Gunther recorded in the classic memoir *Death Be Not Proud,* "That a man of this rank should leave his desk and go to a child's bedside hundreds of miles away on five minutes' notice is sufficient indication of his character."[9]

The boy was brought to the Neurological Institute by ambulance, and Putnam operated the next day, partially excising the mass that proved to be a malignant glioma. A few weeks later, Wilder Penfield came to consult; and propelled by his years of cytological training and research, he pored over the biopsy slides and confirmed the deadly prognosis. For the family of a celebrity as famous and rich as John Gunther, Sr., there were no limits to medical care; but nothing led to sustained improvement, not even a second operation by another highly respected Neurological Institute neurosurgeon, Lester Mount. They contacted Joseph Burchenal at the Memorial Sloan-Kettering cancer hospital across town; he was one of the first to use chemotherapy for cancer, using nitrogen mustard. But that and every other treatment proved fruitless. Periods of improvement did not last, and the boy died on June 30, 1947. Both Putnam and Penfield wrote tender letters of condolence.

The dates have implications for Tracy Putnam. He last saw young Gunther when he visited New York a week before the boy's death. Putnam came from California, where he had been for more than a year after resigning as director of the Neurological Institute.

Notes

1. Oxford Dictionaries Online, 2007: www.oxfordlanguagedictionaries.com

2. Brenner S. Manners, good and bad (Book review of James D. Watson, "Avoid Boring People," Knopf, New York, 2007). Science 2007;318:1245–46.

3. Letter. T. J. Putnam to Meyer Friedman, May 16, 1961. Louise M. Darling Biomedical Library, Center for the Health Sciences, UCLA.

4. Roazen P. Putnam Irmarita, a fine thing for normal people. In How Freud worked. First-hand accounts of patients. Northvale, NJ: Jason Aronson, 1995, pp 167–194.

5. Dies from Vassar fall. Daughter of T. J. Putnam was an honor student. New York Times, September 25, 1947.

6. Letter. T. J. Putnam to Meyer Friedman, May 16, 1961. Louise M. Darling Biomedical Library, Center for the Health Sciences, UCLA.

7. Aird RB. Foundations of modern neurology, a century of progress. New York: Raven Press, 1994, p 16.

8. Letter. T. J. Putnam to Meyer Friedman, May 16, 1961. Louise M. Darling Biomedical Library, Center for the Health Sciences, UCLA.

8a. The National Research Council was a federal agency, forerunner of the National Academy of Sciences and the National Institutes of Health. The Health Insurance Plan (of New York) was one of the first prepaid health insurance plans; this one was based on city employees and is now still in business. The Board of Neurosurgery and the one for Psychiatry and Neurology are national accrediting agencies, setting criteria for training and examination.

9. Gunther J. Death be not proud. A memoir. New York: HarperCollins, 1949.

In 1943, Mr. Charles Cooper was elected President of the Presbyterian Hospital. He at once began a campaign to reduce the medical school to an appendage of the hospital, and made a few menacing gestures in my direction. In 1945, he (presumably) sent me word through Dr. Robert Loeb, who was then being groomed for Professor of Medicine, that I should get rid of all the Jews in my department or resign. I replied that I had no intention of doing either, and that he could not force a professor to resign; but I was wrong about the latter. The faculty meekly fell in line with Cooper's instructions, despite a few scattered protests, and I saw that I should have to leave. This was a source of much distress to me, for with the end of the war I visualized enormous possibilities for scientific advancement. They have not been realized, and the neurological department at Columbia has lapsed back into the state of anxious inactivity which I found there in 1939.

Tracy J. Putnam, letter to Meyer J. Friedman, MD, 1961[1]

6

The Jewish Question and Political Liberalism

George Humphreys, a surgeon at Columbia, stated that both Putnam and his wife were political liberals, which was one cause of Charles Cooper's enmity. Putnam himself attributed the animosity to his refusal to fire the Jewish neurologists on his staff.

The political and religious issues overlap, and there is little documentation in the records of either Columbia University medical school or the Presbyterian Hospital. The Nazi-Soviet Peace Pact was signed on August 23, 1939. The alliance of these archenemies confused supporters on both sides. On September 1, the Germans invaded Poland. American reaction to these events was mixed; the opposing views were not "right" and "left" but rather "pro-" and "anti-interventionalist." The "antis" were sometimes called "isolationists." The differences were based on either a person's views about helping Britain and France, targets of the pact, or the possibility that World War II might commence.[2] People of diverse political persuasions found themselves on the same side with political enemies.

In May 1940, members of the American Association of Scientific Workers, led by Nobel laureates Arthur H. Compton and Robert Mulliken, prepared a "peace resolution" to argue against the "the futility of war."[3] Putnam was one of hundreds of notable biomedical scientists who signed the petition for peace, which appeared on May 20 as a full-page advertisement in the *New York Times*.[4] Eight days later, Walter Cannon (the renowned Harvard physiologist) and Putnam publicly retracted their participation when the Germans invaded

Belgium.[5] The official American position on the Soviet Union changed from hostility to sympathy on June 22, 1941, when Germany actually invaded Russia; the United States then supported the Soviets in their battle against Hitler. With amity replacing enmity, it was not so radical for Putnam to have supported Russian war relief in 1941, as announced on the day Pearl Harbor was bombed and the United States entered the war as an ally of the Soviet Union, England, and France.[6]

The Jewish question may have been related to the political question because it was still evident a decade later, when the author was a resident at the Neurological Institute (1950–1953). That is, the Neurology Service was divided into two parts, East and West. The East Service, headed by Irving H. Pardee and then L. Beverly Chaney, was staffed by society neurologists; Pardee himself married Abby Rockefeller, daughter of John D. Rockefeller. His service did not include any Jewish neurologists.

The West Service was directed by Henry Alsop Riley. At the time, I was a resident in training, and I credited Riley for the presence of the five German-Jewish neurologists who practiced and taught. The first, Alfred Gallinek, came in 1936; but all the others were appointed in the Putnam years: Ernst Herz, Werner Hochstetter, and Joseph Moldaver arrived in 1940 and Maximilian Silbermann in 1944. Two other German neurologists were on the staff. Paul F. A. Hoefer came with Putnam from the Boston City Hospital to establish the electroencephalography unit in the institute; he was neither Jewish nor outspokenly political, and his motivation in leaving Germany was not discussed with his associates. In Fred Mettler's neuroanatomy laboratory, the noted German clinician and neuropathologist Otto Marburg, also a Jewish refugee, arrived in 1939 and had been sponsored by Putnam.[7]

Gallinek was a competent clinician and a kindly clinical teacher. Each of the others made a special contribution. Herz was one of the first to use motion pictures to document the characterization of different movement disorders; Putnam himself was actively interested in Herz's project. Hochstetter was a pioneer in electroshock therapy, and Moldaver was doing early work to develop nerve conduction studies in the differentiation of nerve and muscle diseases. Silbermann was the unofficial and popular neuropsychiatrist for the Metropolitan Opera singers and musicians; one wealthy patient provided him with the only Rolls Royce car in the doctors' parking lot at the Institute.

Since 1919, Otto Marburg (1874–1948) had been director of the Neurological Institute in Vienna. He was a prolific writer of scientific papers and books and had brought with him a superb collection of tissue sections from patients who had had multiple sclerosis. In 1906, he had described an acute form of the disease, now called the "Marburg type." He came to the United States in 1938 and was appointed to the Columbia faculty in 1939, as a clinical professor of neurology, serving until he died. He gained some public fame when he was denied an application to practice neurology without taking an examination, losing an appeal in 1941.[8] An obituary[9] stated that Marburg was granted a license later through the intervention of friends and associates. Marburg was recognized by election to honorary membership in the American Neurological Association.

Another émigré was Hans Hoff (1897–1969), who had been exiled from his native Vienna, where he had been head of the Neurological Institute there. Forced to leave after the German invasion and Anschluss in 1938, he went to Iraq, where he helped to establish neurology in Baghdad. After four years there, he moved to New York and, in 1949, became one of the few refugees to return to the country and city that had treated him so harshly. He welcomed drug therapy, partly because it helped psychiatric patients talk to their therapists. An authority on the treatment of alcoholism, he lived another 20 years.[10]

It seems likely that all of these European neurologists were appointed by Putnam, who was the vice chair of the National Committee for Resettlement of Foreign Physicians,[11] which had been founded in 1939.[12] The executive committee comprised figures from Boston, New Haven, Baltimore, and New York, with others from widespread sites. Despite the humanitarian need, there was considerable opposition to the importation of refugee physisicans.[13] Among the sponsors of the National Commitee for Resettlement were several of Putnam's colleagues from Harvard, including Harvey Cushing and Stanley Cobb.

Between the years 1933 and 1940, Nazi refugees accounted for 3097 of the 5056 immigrant physicians (61.2%). They were not welcomed everywhere but, according to historian Kohler,

Refugee physicians benefited from the fact that New York was home to many victims of the antisemitic and anti-Italian admission policies adopted by American medical schools after World War I. New York was the only state in the country where, because so many had been forced to study medicine abroad, licensure of foreign-trained physicians was never an issue. Also, unlike most other states, New York did not require a virtually unpaid year of internship as a precondition for admission to its examination.[13a]

It is difficult now to backtrack and replay the temper of the times about Jewish faculty and students in the elite universities during the 1930s. The number of Eastern European Jewish students rose progressively. In New York, almost 80% of the students at City College were Jewish, as were 50% at New York University and 40% at Columbia. In New Haven, the Jewish population rose, and so did the number of applications from their sons. In 1911, the first-year class included three Jews; in 1926, there were 27. In the years 1911 to 1914, 5.5% of the undergraduate Yale classes were Jewish; a decade later, the figure was 10.2%. In 1922, Harvard had a class that was 5% Jewish. Both of these leading colleges thereupon set quotas of about 10%.[14]

At the individual level, Harvey Cushing's comments are illustrative. He wrote a recommendation to Phillips Exeter Academy for an applicant who was the son of one of his patients; the young man's surname was "Cohen." He was, Cushing wrote, "an especially attractive and fine-looking youngster with, so far as I can see, little of the Jew about him except his name."

Cushing was fond of his outstanding pupil Leo Davidoff, whom he considered "'a very un-Hebraic Hebrew'—escaped the effects of the ghetto—a Hebrew but a most attractive lad. You will have to have a Hebrew or two on your staff and he is about as good a one and as unhebraic."[15] (A generation later, the Albert Einstein College of Medicine opened with Jewish sponsorship and an affiliation with Yeshiva University. The "unhebraic" Davidoff was the first chief of Neurosurgery there.)

In 1925, Jewish quotas were coming along, and Cushing decided he did not want more than three on the Brigham staff at the same time. "I have no objection to Hebrews but I do not like too many of them all at once." Cushing, however, did agree to a request from Ernest Sachs to serve on the New York–based Emergency Committee in Aid of Foreign Displaced Physicians and to have his name on the letterhead. In July 1939, he agreed to be one of the representatives from Connecticut on the National Committee for the Resettlement of Foreign Physicians, formed under the leadership of chair David Edsall.

Biased views were similarly held by other leading figures, including Nicholas Murray Butler, Nobel laureate for peace and president of Columbia University from 1902 to 1945. He refused to give an honorary degree to the great cell biologist Jacques Loeb or to support that scholar's nomination for membership in the prestigious Century Club in New York City, on the simple grounds that Loeb was Jewish. More to the point, President Butler devised numerous ways to reduce Jewish enrollment at Columbia, including an ill-fated and transient establishment of a separate campus for Jewish students in Brooklyn, far from the uptown campus in Manhattan.[16]

Determining who was Jewish could be complicated, as illustrated by Jacques Loeb and his son, Robert F. Loeb. Other family members were also illustrious biomedical scientists.

Jacques Loeb (1859–1924)

Jacques Loeb was nominated for a Nobel Prize more than 70 times but never made it to Stockholm. He was nevertheless friends with contemporary laureates Albert Einstein and Lord Rutherford.

Trained in Germany, he came to the University of Chicago in 1892. After a decade there, he moved to the University of California, and in 1910 he went to the Rockefeller Institute, where he remained until he died. In summers he worked on invertebrates at the Marine Biological Laboratory and found that he could induce embryonic development by chemical changes in the water and without sperm, called "artificial parthenogenesis."

As noted earlier, Jacques Loeb was at least twice snubbed and denied Columbia honors by Nicholas Murray Butler for being Jewish. In fact, however, he was born in Alsace in 1859; a "freethinking and non-observant Jew,"[15a] he considered himself an agnostic atheist who loathed Prussians and militarism; he spoke French rather than German. He was orphaned at age 16 and had a younger brother, age 6. They were taken in by an uncle, Harry Breslau, also a nonpracticing Jew and a professor of history at the University of Berlin. One of Harry's daughters married Albert Schweitzer, another of many intermarriages in the family.

Jacques had a brother, Leo, who was chair of Pathology at Washington University in St. Louis. Jacques' two sons, Leonard and Robert, were equally illustrious. Leonard was a professor of physics at the University of California.[17]

The second son, Robert F. Loeb, was a great medical professor and consummate teacher at Columbia University.

Jacques also studied the effects of brain injuries on sensation and was an early user of the electroencephalogram, but he was primarily a biochemist who studied proteins; he became even more famous by way of Paul de Kruif, author of *Microbe Hunters*, a hugely popular book about historical figures such as Louis Pasteur and Robert Koch, who contributed to modern concepts of infectious disease. For generations, that book lured many young readers to careers in medical practice or research. de Kruif himself had been successful in research at the Rockefeller Institute and there became an admirer of Jacques Loeb. Torn between possible careers in either medical research or literature, de Kruif became the model for the protagonist in *Arrowsmith*, a novel by Sinclair Lewis. The powerful character of Gottlieb, a dedicated investigator in that story, was based on Loeb.[18] Lewis won the Nobel prize for Literature in 1930, five years after publication of *Arrowsmith*.

Walter W. Palmer (1882–1950)

Walter Palmer was appointed chief of Medicine at Presbyterian Hospital in 1921 and served for 26 years.[19] With the limited activity of Frederick Tilney, Palmer was given additional responsibilities as director of the Neurological Institute from 1935 to 1939, when Putnam arrived. Busy himself as chief of Medicine, Palmer appointed Robert F. Loeb to be associate medical director in 1938; Loeb maintained an appointment in the institute until 1946.

Palmer himself was a historic figure in academic medicine.[20,21] As an undergraduate at Amherst College, he was a star football player; he then went to Harvard Medical School. He interned at the Massachusetts General Hospital and was set to go into practice in 1912 but became interested in learning why kidney disease caused retention of water in the legs. This was his entry into clinical chemistry. After three years at the Massachusetts General Hospital, he moved to the Rockefeller Institute in New York in 1915; but two years later, he moved to Columbia as acting director of the Medical Service during World War I. Again, he did not last long in that position and moved to Johns Hopkins in 1919. Two years later, he was invited to be chief at four leading universities. His popularity was based on his scientific record, his personality, and his interest in the organization of medicine. He was committed to the idea of a full-time system with teacher–investigator-clinicians as the core. He was therefore attracted to the new Columbia-

Presbyterian Medical Center and brought with him from Johns Hopkins Robert Loeb and Dana Atchley, who had similar research interests. They were all analyzing and treating the chemical abnormalities of diabetic crisis and trying to determine the role of iodine in thyroid disease. He was an important editor of journals and textbooks. "He was a big man both in physique and spirit."[21a]

Robert F. Loeb (1895–1973)

Robert, the son of Jacques Loeb, was a commanding presence. He was one of the first to do "clinical investigation," applying biochemical methods to study the changes in blood constituents of patients in crisis from uncontrolled diabetes mellitus or the adrenal failure of Addison disease, which was common and serious in those days of widespread tuberculosis. After World War II, antibiotics limited the spread of tuberculosis and adrenal failure became rare. In doing those investigations, Robert Loeb followed his teacher, Walter Palmer, in changing medical research and the treatment of human diseases from the merely observational and descriptive to human experimentation and biochemistry. He was also a consummate bedside teacher of medical students and a warm-hearted physician. He became a major force in influencing the direction of medical research nationally and was the editor of an influential textbook of medicine.[22]

Robert Loeb was chair of Medicine at Columbia from 1947 to 1960. His appointment again raised the Jewish question. Because Robert's father, Jacques, was considered Jewish, it was assumed by many that the sons were, too. However, that was not true.

Robert's mother, Jacques' wife, was Ann Louise Leonard, who had been reared as a Congregationalist in East Hampton, Massachusetts, near Smith College. She was one of three sisters sent by G. Stanley Hall (president of Clark College) to Europe for graduate school, and all three gained PhD degrees. One of them became president of Bryn Mawr College. She was probably Marion E. Park (1875–1960), who was president of the college from 1922 to 1942; she would have been 47 years old when appointed to that office. To round out that illustrious sibship, Robert Loeb's maternal uncle was E. A. Park, head of Pediatrics at Yale.

Throughout adolescence and adult life, Loeb went to church regularly; he had a mixed religious background himself, and his Jewish ancestors were not observant in any way. Loeb never considered himself Jewish. He married Emily Nichols, who came from an impoverished family, started as a technician in his laboratory at Johns Hopkins in 1922, and went to medical school there (1929–1933).

Milton C. Winternitz (1885–1959): The Yale System of Medical Education

Meanwhile, at Yale Medical School the quota system was instituted by a Jewish dean, Milton C. Winternitz.[23] He was of Czechoslovakian heritage and half Jewish. Dean from 1920 until 1935, Winternitz is universally regarded as having had a major impact on the school, transforming it from second rate to top tier by recruiting outstanding faculty and adding buildings. Perhaps his most enduring contribution was a philosophy of education. He respected medical students as intelligent and mature enough to set their own goals and select their own courses without the frequent examinations that had been considered necessary before he arrived. Under his leadership, examinations were "offered" as teaching aids but not "given" as tests of learning. He wanted medical students to be treated as if they were mature graduate students, including a requirement for a research thesis. This "Yale system" endures.[24]

Yet, there was a dark side to Winternitz. He seemed to spend his life trying to escape his Jewish heritage. He had been a professor of pathology at Johns Hopkins. His chief, William Henry Welch, told him that he had no future there because he was young and "for other reasons as well,"[24] presumably including being Jewish. The opportunity to work at Yale was a godsend.

As part of his social orientation, Winternitz admired high society and acted on his aspirations by marrying Helen Watson. She was the daughter of Thomas A. Watson, partner of Alexander Graham Bell and immortalized by the first words ever spoken on a telephone. On March 10, 1876, Bell recorded the episode in his notebook, commanding: "Mr. Watson—come here—I want to see you." The next year the pair founded the Bell Telephone Company and were forever financially secure—undoubtedly, that included Helen Watson.

When Winternitz arrived in New Haven, he was surprised to find it difficult for Jews to buy a house in a fashionable neighborhood, even for a newly appointed dean. And he had difficulty being admitted to the Graduate Club, a faculty dining room. He succeeded because letters were written on his behalf by a former president of the United States, William Howard Taft, and by the president of Yale, Arthur T. Hadley.[25] Instead of being provoked to oppose these policies, Winternitz promulgated a simple formula for balance in the student body: "Never admit more than 5 Jews, take only 2 Italian Catholics, and take no blacks at all."[26]

Comments about his personality conveyed his Janus-faced attributes:

"He was a genius, a great teacher, one of the best ever."
"He was a bastard and power-hungry."
(Just over five feet tall) "he had the typical Napoleon complex that goes with short stature."
"He had virtually everybody scared of him, including faculty."
"He was at times brutal and sadistic and could massacre students."
"Students revered him and feared him and loved him and hated him."
"He carried with him a sense of greatness."[27]

On being Jewish:

"He didn't want to have anything to do with his Judaism."
"He did not like being Jewish. He did not like the characteristics that Jews had."
"He was the most anti-Semitic person on the faculty. He hated them."[28]

But Jewishness was not an issue in his domineering teaching behavior:

"He taught students to accept nothing just because someone had told it to them. . . . He was brutal in humiliating students in class and several students suffered greatly under this kind of teaching. They almost didn't know their first names by the end of the class period."[29]

Writer Susan Cheever's mother was Mary Winternitz, daughter of Milton and wife of John Cheever. Susan wrote that her grandfather was "a short man with a tyrannical manner, an intense charm that could make you feel you were the only person in the world—and a raging temper that could make you wish you weren't."[30]

Spiro and Norton[31] gave a balanced assessment of Winternitz but tended to attribute his anti-Semitism to pervasiveness in medical schools of the time.

Anti-Semitism in universities and medical schools started to decline after World War II, when veterans returned and attended college as beneficiaries of the G.I. Bill. Also, the success of organized science during that war led to an increasing role of the federal government in funding research, including biomedical research. German immigrants contributed prominently to the rise of biochemistry as well as atomic physics. In the 1950s, the National Institutes of Health began awarding research grants that included money for indirect costs—construction, record keeping, maintenance of laboratories and buildings. Universities competed for scientific talent in recruiting faculty, ignoring the old restrictions.[32] Grants were based on merit, not on the social status of the investigator; and Jewish quotas promptly disappeared from admissions policies. Nevertheless, as late as 1967, I became the first openly Jewish chair of a clinical department at the University of Pennsylvania and, 6 years after that, I became the first Jewish clinical chair at Columbia (since Elsberg retired in 1939). The assimilated, anglicized surname may have made those appointments easier for trustees.

Both institutions had distinguished chairs with Jewish heritage. At Penn, Isidor S. Ravdin had what might seem to be "a Jewish name," but he was never clearly identified as Jewish. His father came to the United States from Latvia, but being Jewish was not mentioned in either an oral history of Ravdin by Professor Saul Benison at Columbia University or a long interview with Lawrence K. Altman of the *New York Times*. One recorded quotation implied erroneously that Ravdin was Jewish:

> Oral histories also help document patterns of exclusion from the "invisible colleges" within academic medicine. Henry Shenkin recalled his dawning awareness, while working as a young neurosurgeon at the University of Pennsylvania, that his ascent up the academic ladder had stalled because he was Jewish. His WASP mentor, Francis Grant explained, "We did it for Rav (Isidor Ravdin, a prominent Jewish surgeon at Penn), but I don't think we can do it for you."[33]

Ravdin's father may have been Jewish but his mother was not and the great surgeon apparently never considered himself Jewish. In that respect, he was like Robert Loeb.

Under the influence of his mother, Robert himself was a regular churchgoer from childhood and became a leader of his church. He never considered himself Jewish. Unlike Milton Winternitz, however, Robert Loeb sponsored the careers of many famous Jewish investigators and professors, including David Seegal, Irving London, Irving Goldberg, Paul Marks, Alfred Gellhorn, and Daniel Kimberg. Winternitz had been active in the 1920s and 1930s; Loeb was chair in the 1940s and 1950s. The times were certainly different, but Loeb and Winternitz had completely different attitudes about Jewish faculty members.

Perhaps it comes down to one question: Who was Jewish in those days? Franz Boas, the great anthropologist, wrote to his son, "You will have to face the fact that in spite of your having only one-quarter Jewish blood, you will probably be classed with Jews—because of me."[34]

Putnam and Jewish Faculty Members at Columbia

Back now to Tracy Putnam. His problems, Jewish and otherwise, led to a confrontation primarily with Presbyterian Hospital, not Columbia University. Views of the hospital's leaders are not in available records, only Putnam's. But the views Putnam attributed to Presbyterian Hospital President Charles Cooper were compatible with common observation about the paucity of an openly Jewish chief of a clinical service until 1973 (when I became chair of Neurology).

Putnam had other Jewish faculty members in his department; Elvin Kabat, David Nachmansohn, and Harry Grundfest came in Putnam's effort to introduce basic scientific research and clinical investigation. Their stories follow in Chapter 7.

An independent view of the Putnam affair was provided by George Humphreys, who chaired the Surgical Service from 1946 to 1969. Humphreys was one of the first cardiothoracic surgeons in New York City. He followed the surgical eminence of Allen Whipple and became eminent himself.[35]

Humphreys became chief of surgery just when Putnam was formally in trouble. Humphreys noted two problems. First, Putnam had tenure, which guaranteed his title and appointment at the university.

By virtue of the Alliance agreement between Columbia University and the Presbyterian Hospital, the university nominates faculty first and recommends candidates to the hospital for endorsement. That policy is still in place, although it has been challenged on occasions when hospital officials wanted to appoint a physician without going through university rigmarole. The case of Tracy Putnam was the obverse, getting rid of a service director.

But, according to Humphreys, the powerful figure who pushed for Putnam's dismissal was the president of the hospital, Charles Cooper. As recorded in a 1990 interview with Michael Myers, this was Humphreys's version:

Dr. Tracy Putnam didn't resign. His job was eliminated. The Putnam episode happened the same year that I came in as Chief. In fact, the decisions had been made before I became Chief as far as the chronology is concerned. There used to be a Director of the Neurological Institute, and Putnam, who was a surgeon, Tracy Putnam, a very well known surgeon, was the Director of the Neurological Institute. He and his wife, particularly, were patients of mine, so I knew them quite well. She, especially, was very liberal-minded. Charlie Cooper was not. He considered that Tracy Putnam was a "pinko." Mr. Cooper didn't like him because he was eccentric in some ways. He certainly didn't fit the role of an administrator that Mr. Cooper had in mind. And so he decided that he should be fired.

However, it was not the prerogative of the Hospital to fire. He went to the Dean and gave him his opinion of Dr. Putnam. And then—I'm not sure, because this preceded when I was really in on things, what happened, except that I think the Dean said he could not fire a member of the faculty with tenure—he had to be discharged for reasons. And there had to be a hearing held by the faculty. This didn't satisfy Mr. Cooper at all. This wasn't the way they did things in business. And so he went to the Medical Board—this is all hearsay, but I think it is pretty accurate hearsay—and talked to the Executive Committee of the Medical Board, to the effect that the situation in the Neurological Institute was untenable, and he thought it was anomalous anyway to have a single director of the Neurological Institute which combined Neurology and Neurosurgery, which didn't fit, and he suggested that the Medical Board decide to eliminate the job of Director of the Neurological Institute and appoint, instead, a professor of neurosurgery and a professor of neurology. So the Medical Board concurred. They were all about to step out, they were tired of administrative problems; it made sense to them; Tracy Putnam wasn't their favorite, anyway. So they went along with Mr. Cooper and eliminated the job. Dr. Putnam was informed then that his position as Director of the Neurological Institute was terminated. But he said, "How about my tenured professorship of neurosurgery?"

"Well, that's up to the faculty."

This was just about the time I came in, in 1946. I remember Tracy coming to me and saying, "Look, this is intolerable. We can't have the administration of the Hospital making academic decisions of this sort. I'm going to stand on my record and not resign." Putnam resigned in '47, the next year.

I suppose technically he may have had to send in a letter of resignation since the job no longer existed. But he certainly didn't resign his professorship, because I remember very well—this was when I was just a new Chief of Service—we had a faculty meeting which Dean Rappleye called, and Putnam came with a big stack of books that he had published as chief of Neurosurgery—he'd done some very outstanding work. He defended himself so well that the faculty refused to vote him out. They said, "No. He has tenure, and there is no reason he should be terminated." So he did not resign his professorship, although he subsequently accepted a professorship in California, and for a while, anomalously, he retained his professorship here while he went to California. Then, I guess, eventually, he did resign—I don't know, his professorship, because he was happy in California.[36]

Notes

1. Letter. T. J. Putnam to Meyer Friedman, May 16, 1961. Louise M. Darling Biomedical Library, Center for the Health Sciences, UCLA.
2. Doenecke JD. Explaining the antiwar movement, 1939–1941; the next assignment. J Libertarian Studies 1986;8:139–162.
3. Mulliken RS. The peace resolution of the American Association of Scientific Workers. Science 1940;91(May):432.

4. 500 Scientists ask US to avoid war. Send petition to president saying progress depends on remaining at peace. New York Times, May 20, 1940. Tracy J. Putnam was named as one of the most prominent signers, which included Nobel laureates Arthur C. Compton, physicist at the University of Chicago, and Dean George H. Whipple, of the University of Rochester. The petition was forwarded by Anton J. Carlson, president of the Association of Scientific Workers and professor of physiology at the University of Chicago.

5. 2 scientists recant recent peace move: W. B. Cannon and T. J. Putnam have had "change of heart." New York Times, May 28, 1940. Cannon was quoted: "Never for a moment did I understand that the resolution meant peace at any price. . . . The resolution was presented to me before recent events. The phrase 'reasonable program' now has a different meaning than it had a few weeks ago. We know now there is a powerful wild beast on a rampage in the world. . . ." Putnam "declared that the invasion of the Netherlands, Belgium, and France had brought about his change of heart."

6. Physicians to help Russia. New York Times, December 7, 1941. Tracy J. Putnam of the Neurological Institute and Dr. Eugene H. Pool of the Hospital for Ruptured and Crippled, named as co-chairmen of a physicians' coordination committee of Russian War Relief, was announced yesterday at the society's headquarters. The group will coordinate the work of 16 hospitals and physicians' committees of Russian War Relief.

7. G. Glaser, founding chair of Neurology at Yale University School of Medicine, was deferred from military service because he was a resident with Putnam in the years 1943–1946, followed by military hospitals of the U.S. Army for two years. He then had a research position in electroencephalography at the Neurological Institute until 1952. Interview with the author, March 10, 2006.

8. Dr. Marburg loses suit for a license. Court of Appeals upholds state in rejecting Austrian. New York Times, July 30, 1941.

9. Dr. Otto Marburg, a neurologist, 74. Professor at Columbia dead—consultant at hospitals here fled from Nazis in 1938. New York Times, June 14, 1948.

10. Culebras A. International newsletter. The sojourn of Professor Hans Hoff in Iraq. Neurology 2003;60:23A–24A.

11. Edsall DL, Putnam TJ. The émigré physician in America, 1941. JAMA 1941;117:1881–1888.

12. Edsall DL. The émigré physician in American medicine. JAMA 1940; 114:1068–1072.

13. Kohler ED. Relicensing Central European refugee physicians in the United States, 1933–1945. Museum of Tolerance Online Multimedia Learning Center, Simon Wiesenthal Center Annual, L.A. 6, 1989, pp 3–32.

13a. Ibid.

14. Oren DA. Joining the club. A history of Jews and Yale. New Haven, CT: Yale University Press, 1985, pp 40–63.

15. Bliss M. Harvey Cushing. A life in surgery. New York: Oxford University Press, 2005, pp 500–507.

15a. Bliss M. Harvey Cushing. A life in surgery. pp —

16. Rosenthal M. Nicholas miraculous. The amazing career of the redoubtable Dr. Nicholas Murray Butler. New York: Farrar, Straus and Giroux, 2006, pp 228, 333.

17. Comments about the Loeb family are based on notes from an interview by the author with John Loeb, MD, professor emeritus of medicine at Columbia University, December 18, 2006.

18. Fangerau HM. The novel *Arrowsmith,* Paul de Kruif (1890–1971) and Jacques Loeb (1859–1924): a literary portrait of "medical science." Medical Humanities 2006;32:82–87.

19. Lamb AR. The Presbyterian Hospital and the Columbia-Presbyterian Medical Center, 1868–1943. New York: Columbia University Press, 1955, p 167.

20. Walter W. Palmer, MD. A minute of the Medical Board of the Presbyterian Hospital in the city of New York, November 21, 1950. Archives and Special Collections, A. C. Long Health Science Library, Columbia University.

21. Dr. W. W. Palmer, physician, is dead. New York Times, October 29, 1950.

21a. Comments about the Loeb family are based on notes from an interview by the author with John Loeb, MD, professor emeritus of medicine at Columbia University, December 18, 2006.

21b. Walter W. Palmer, A minute.

22. Bearn AG. Robert Frederick Loeb (March 14, 1895–October 21, 1973). Biographical memoirs, vol 49. Washington DC: National Academies Press, 1978, pp 149–183.

23. Oren, Joining the club, 136–158.

24. Spiro H, Norton PW. Dean Milton C. Winternitz at Yale. Perspect Biol Med 2003;46:403–412.

25. Oren, Joining the club, 143.

26. Ibid., 148.

27. Ibid., 141.

28. Ibid., 142.

29. Ibid.

30. Burrow GN. A history of Yale's school of medicine. Passing torches to others. New Haven, CT: Yale University Press, p 97.

31. Spiro, Norton, Dean Milton C. Winternitz.

32. Ludmerer KM. Time to heal. American medical education from the turn of the century to the era of managed care. Oxford: Oxford University Press, 1999, p 145.

33. Tomes N. Oral history in the history of medicine. J Am Hist 1991; 78:607–617.

34. Cited in Brickman JC. "Medical McCarthyism": The physicians forum and the cold war. J Hist Med Allied Sci 1994;49:380–418 (footnote 9, p 382).

35. Herter FP, Jaretzki A III, Forde K. The birth of cardiothoracic surgery at Columbia: George H. Humphreys II (1946–1969). In A proud heritage. An informal history of surgery at Columbia. New York: Department of Surgery, College of Physicians & Surgeons, Columbia University, 2003.

36. The quotation is from the transcript of a conversation between George Humphreys and Michael Meyer. The interview took place in about 1990, as preparation for a history of the College of Physicians & Surgeons of Columbia University.

Within a year, Dr. Putnam hopes to use the magnificent resources of the Institute for large scale clinical work in 1) multiple sclerosis, a mysterious nerve-crippling disease, probably twice as prevalent as infantile paralysis; 2) paralysis agitans, a lingering, incurable shaking palsy; 3) epilepsy (known to modernists as "convulsions"). Meanwhile, within the cheerful green walls of the Institute, turbanned patients continue to wheel their chairs through sunny wards, as 100 experimenters work on problems such as mirror-writing, abnormalities of the senses, hydrocephalus (water-on-the-brain), brain physiology and anatomy.

Medicine. Bread-&-Butter Brains.
Time, October 16, 1939[1]

7

Putnam Establishes Basic and Translational Neuroscience at Columbia

Before Putnam

Before Putnam arrived in 1939, there had been some neuroscience research at Columbia but not much. However, Putnam's faculty appointments, conversely, led in a direct line from Harry Grundfest to Nobelists Eric Kandel and Richard Axel, from Fred Mettler to Malcolm Carpenter in neuroanatomy, from Putnam's own work on multiple sclerosis to Elvin Kabat and, later, to patient-oriented research in the Department of Neurology for a panoply of human disorders: epilepsy, neuromuscular disease, Mendelian and mitochondrial genetic diseases, epidemiological studies of dementia and cerebrovascular disease, and child neurology. For a while, however, all neuroscience in the entire university was in Putnam's Department of Neurology.

Before Putnam, Tilney and Riley had produced *The Brain from Ape to Man* and a widely used anatomical atlas of the human brain and spinal cord,[1a] but the Department of Neurology had not carried out much disease-oriented research, in stark contrast to the epilepsy progress seen at the Boston City Hospital.

Before Putnam, the work of one basic scientist could qualify for the designation "neuroscience." K. S. Cole (1900–1984) was appointed assistant professor of Physiology in 1929 at age 29.[2] He spent summers at the Marine Biological Laboratories in Woods Hole, Massachusetts. There, in 1936, he was introduced to the giant axon of the squid by its discoverer, J. Z. Young. That unique nerve became the focus of intense international research. That same year, he met and joined Howard J. Curtis in a collaboration that lasted six years, ending when Curtis moved to Johns Hopkins. He had become increasingly interested in how nerves transmit electrical impulses. Cole left Columbia in 1942, moving to the University of Chicago to be in charge of the biological effects of radiation released by the atomic bomb, which was under development there. It is said that he and nuclear physicist Leo Szilard voted against using the bomb. Cole had a distinguished career, which included development of the "voltage clamp," a technique that became important in later work on nerve transmission, that is, how the action potential moves down from the cell body to control the activity of a muscle or gland.

Before Putnam, another early scientist did focus on a human disease. Abner Wolf (1902–1983), a graduate of Columbia's College of Physicians and Surgeons, was trained in Germany by two famous neuropathologists: L. Pick, who had a disease named after him (now considered a form of frontotemporal dementia), and A. Jakob, who gained another eponym, Creutzfeldt-Jakob disease. Wolf joined the Columbia faculty in 1931 and was appointed to the staff of the Neurological Institute in 1933. He attained the rank of professor of Neuropathology in 1951 and was ultimately an international leader in his field.

In 1937, Wolf and his student, David Cowen, described a then newly recognized brain disease of infants, toxoplasmosis, the same disease that later became a scourge of people with AIDS. He continued to write about clinical and experimental studies of that infection for almost a decade and later joined Elvin Kabat, the first of Putnam's basic science appointees, to elucidate the pathogenesis of multiple sclerosis with an animal model of "autoimmune experimental encephalomyelitis."

Elvin A. Kabat (1914–2000)

Elvin A. Kabat was a wunderkind. He graduated from City College in New York in 1932 at age 18, in the depth of the Depression. To pay the costs of graduate school, he worked in the laboratory of Michael Heidelberger, an association that was to be superlatively productive for the rest of their long lives.[3] In 1937, Heidelberger sent Kabat to Sweden for collaborative work with Arne Tiselius, a physical chemist and pioneer in studying the properties of proteins. They hit gold with one of Kabat's first studies, identifying antibodies as gamma globulins. Heidelberger and Kabat became famous for quantitative studies of the interaction of antibodies and foreign proteins to protect humans against bacteria and viruses. Knowing that the antibodies were found in the gamma globulin fraction of proteins in the blood was a major step forward in purifying the protective antibodies for the studies they did subsequently.

Kabat returned to New York in 1938 as an instructor at Cornell Medical College. There, he worked on tumor viruses, and in 1940 Putnam asked Michael Heidelberger to recommend an immunochemist to work on multiple sclerosis. Kabat wrote, "He invited me for an interview and said that he understood I had lots of my own ideas and that I could work on whatever I wished, except he hoped I would not discriminate against neurological problems."[3a] Putnam provided two rooms in the Neurological Institute for laboratories, and funds were available for a technician. His salary would increase from $2400 at Cornell to $3600 at Columbia. Still, he wavered until Robert F. Loeb, a friend from the Marine Biological Laboratory in Woods Hole, Massachusetts, gave strong encouragement.

Immediately on starting to work in June 1941, Kabat began to study the proteins of cerebrospinal fluid (CSF) with the Tiselius electrophoresis apparatus. Kabat struck gold once again. He could identify gamma globulins in the CSF and noted abnormally increased amounts in samples from patients with multiple sclerosis or neurosyphilis. The results were sufficiently consistent that the test became widely used in the diagnosis of both diseases. The new CSF test rapidly replaced a much less specific one called the "cephalin flocculation test"[4] on the CSF, but no one knew why the appearance of "flocculation," a cloudiness of insoluble material, in the test fluid was often an indicator that the disease was present. Kabat showed that flocculation was the result of high CSF concentrations of gamma globulins. These results were important theoretically—pointing to an immunological disorder in multiple sclerosis—and practically because the CSF test promptly became important in the diagnosis of both multiple sclerosis and neurosyphilis. But it took another 20 years to stop doing the flocculation test at the Neurological Institute because tradition favored it.

During World War II, Kabat worked on diverse subjects, wrote a book on immunochemistry that became a classic, and worked on germ warfare several days a month at Fort Detrick in Maryland. After the war he focused on two totally different projects, the chemistry of blood group substances and "acute disseminated encephalomyelitis," or ADEM, an experimental disease that was considered a model of multiple sclerosis. In both multiple sclerosis and ADEM, the pathological reaction seemed to be loss of the protective myelin sheath on nerve fibers called "myelin" and likened to insulation on an electric wire. This "demyelination" had first been discovered by Thomas Rivers and Francis Schwentker at the University of Michigan, who were trying to find the cause of "paralytic accidents," which followed use of the Pasteur treatment for rabies, a course of injections routinely given to children or adults who had been bitten by a dog or some wild animal. Without the vaccine, rabies was virtually always fatal and the vaccine spared lives, so there was a dilemma: save the life or risk the adverse effects. It would be important to understand how the paralytic accidents came about.

The vaccine was produced by growing the rabies virus in the spinal cord of animals, and the investigators found the culprit when they injected monkeys with suspensions of spinal cord (without the virus). In humans who had died of a neurological reaction to the vaccine and in experimental monkeys, "demyelinating" lesions like those of multiple sclerosis were found in the brain and spinal cord. One problem for the investigators was the large number of injections needed to induce the experimental disease and, therefore, the long time it took before symptoms appeared.

Kabat joined with Ada Bezer and Abner Wolf to see if an adjuvant could enhance and accelerate the immunization process; the adjuvant was a mixture of lipids that had a nonspecific enhancing effect on chronic immunological reactions. It did the job, reducing the number of injections to two or three. Their results were published in 1946 and induced the newly formed National Multiple Sclerosis Society to give Kabat its very first research grant.

Kabat's interest in multiple sclerosis continued for 30 years because his laboratory in the Neurological Institute carried out the diagnostic test for CSF gamma globulin until his immunoprecipitation method was replaced by automated techniques. In those later years, his chief of Neurology was H. Houston Merritt, who came to Kabat's support when needed.

After the war, Abner Wolf had become a part-time consultant at the Veterans' Administration (VA) Hospital, and Kabat joined him in research on the intracellular location of enzymes. To facilitate his participation, he was also made a part-time consultant; he was now a part-time employee of the federal government. There followed the executive order of President Harry S. Truman, which initiated loyalty investigations. James B. Sumner, a biochemist, went to the FBI and told them Kabat had been a communist while they were in Sweden together. The consequences had become familiar—FBI investigation, dismissal by the VA loyalty board, reversal and reinstatement by a higher board, continued harassment, resignation from the VA, and termination of his histochemical research program. Kabat also lost his passport in a program to limit travel for people with leftist political views; the program was ultimately deemed unconstitutional and eliminated.

Throughout this period, Kabat made progress in determining the immunological basis of severe allergic reactions to dextran, which had been developed as a plasma expander or blood plasma substitute, to help treat circulatory shock after blood loss. In 1953, however, long after Putnam's departure, Kabat's research on multiple sclerosis ran into political trouble. His application for renewal of the National Institutes of Health (NIH) grant for the study of allergic encephalomyelitis in monkeys was turned down, for reasons that had nothing to do with science. The letter stated only that the "request falls in the group of applications for which grants cannot be made."[5]

When this had happened to Nobel laureate Linus Pauling, he had named someone else in his group as responsible investigator, but Kabat would not accept that compromise. He refused to have anything to do with the Public Health Service, federal home of the NIH. The description in his autobiography confirms a story that was widely circulated at Columbia. Kabat received a letter from the secretary of Health and Human Services, informing him that the NIH would like Kabat to host a visit to his laboratory by a professor from Puerto Rico. Kabat replied that he would surely welcome the visit but not under the sponsorship of the NIH, which had canceled his grant. In contrast, Kabat would welcome the visitor if he came under the auspices of the U.S. Army, the Navy, or the Atomic Energy Commission, all of which had continued their grants without interruption.

But the harm had been done. Wolf, Bezer, and Kabat "had to kill off the only monkey colony in the world then being devoted to the multiple sclerosis problem." That ended his team's work on the disease, "just when (they) were planning to isolate the encephalitic antigen. Fortunately, this problem was taken up by many others."[5a]

Kabat's later work included the amino acid sequences in antibodies and heavy-duty physical chemistry estimates of the structure of the combining site of antibodies that interact and bind the antigens; his data became essential elements of modern immunology.[6] He was awarded the National Medal of Science in 1991, the highest award given to scientists in the United States. He died of a lymphoma in 2000, a disorder possibly related to his self-experimentation, injecting himself many, many times with diverse antigens so that he could himself be an antibody factory.[7] Obituaries were naturally laudatory.[8,9] One named him the "founder of modern immunochemistry" and noted that one of Kabat's students, Baruj Benacerraf, won the Nobel Prize, "an honor that eluded both Kabat and his mentor, Heidelberger, although many believed that their contributions amply merited this contribution."[10]

Harry Grundfest (1904–1983)

Harry Grundfest was born in Russia; came to the United States at age 9; went to high school in Kearny, New Jersey; and then enrolled at Columbia College. In the mandatory course on western civilization, he learned that a professor at Columbia was doing genetics research on fruit flies. As a result, he enrolled in a seminar led by Nobelist Thomas Hunt Morgan. In 1925 he started graduate training with Selig Hecht, a pioneer in the study of color vision, receiving a PhD in 1930. He then moved to the University of Pennsylvania and befriended Deltev Bronk. He returned to New York to work at Cornell medical school with Herbert Gasser, a neurophysiologist and another Nobel laureate. Gasser was studying how nerves transmit information. In 1935, Bronk moved to the Rockefeller Institute and recruited Gasser and Grundfest to join him, and Grundfest remained there until 1945. Grundfest published six papers together with Gasser in the years 1936–1939. Among their other achievements, they clarified the relationship between the diameter of a nerve and the speed of conduction. In 1944, Gasser won a Nobel Prize with Joseph Erlanger, but Grundfest was not named with them.

That period covered World War II. Grundfest's curriculum vitae stated simply that, in the years 1943–1945, he was on leave of absence as a "Senior Physiologist, Climatic Research Unit, Fort Monmouth Signal Laboratory." His research was concerned with adaptation to cold weather, including clothing and heaters. Several of his reports were labeled "restricted" by the Army. (Their ultimate fate has not been tracked.) That seemingly bland assignment was to have fateful consequences.

Grundfest left the military in 1945, worked briefly on the effects of gunshot wounds, and was then recruited to Columbia by Tracy Putnam. His first academic title was that of research associate in neurology, which involved no teaching, only research. He had his own laboratory for neurophysiology and directed the Peripheral Nerve Study Center to follow the evolution of gunshot wounds in veterans. He was promoted to assistant professor in 1947 and associate professor in 1949. He gradually created a sophisticated electronic laboratory in the dilapidated two-story Kramer Building, named for the surgical supply store on the first floor, a block away from the medical center.

Each summer, Grundfest moved tons of equipment as well as several investigators and graduate students to the Marine Biological Laboratory at Woods Hole, Massachusetts. There, they worked for four months, often using the giant axon of the squid to figure out how a nerve carries an electrical impulse. Similarly, he elucidated the process of "excitation–contraction coupling," how electrical current generated in the surface membranes of muscle can cause the muscle to contract and generate a force. In addition to the science itself, his lab was an active training ground. Among his students were Eric Kandel, Dominic Purpura, Michael Bennett, George Pappas, and John Reuben. His trainees included many from Japan, and he was given the highest honor of that government for foreigners, the Order of the Rising Sun, as well as awards from numerous other countries. He was elected to the National Academy of Sciences in the United States and to the equally prestigious Physiology Society of London. His published scientific papers came to more than 500.

In the early 1950s he chaired the medical advisory board of the Hebrew University in Israel, but that was terminated in 1954 because, like Elvin Kabat, Grundfest was denied a passport. That came about when he was propelled to national prominence by Senator Joseph McCarthy in November 1953. Grundfest was called before the Senate Permanent Subcommittee on Investigations. According to a newspaper report,[11] Grundfest "denied ever engaging in espionage" but refused to answer questions about communists. A trustee of Columbia University called for his dismissal for taking the Fifth Amendment. Grundfest's NIH grants, like those of Kabat, were closed. He had done research in close association with David Nachmansohn, but they ended disagreeing about the role of acetylcholine in nerve conduction; after the McCarthy affair, Nachmansohn denied Grundfest access to heavy equipment they had obtained on grants for shared research.

John Reuben wrote, "Without the strong backing of Houston Merritt, who was both chairperson of the Department of Neurology and dean of the medical school at the time, he might not have survived the unwarranted persecution with his university position intact."[12] Merritt did not agree with Grundfest's political views but ignored or resisted Columbia University alumni and others who called for his dismissal.

During the years 1954–1961, Dominic Purpura was in the laboratory, and they worked together on cortical functions of the cerebrum and cerebellum. They concluded that transmission between neurons in the central nervous system is fundamentally similar to the much more readily studied transmission from nerve to muscle. Extrapolated, that means the complexity of the brain and spinal cord does not imply special properties of the neurons but is the result of the multiplicity of cells and interconnections. They also discovered that gamma-aminobutyric acid (GABA) is an inhibitory transmitter. Purpura later became head of Neuroscience at the Albert Einstein College of Medicine and dean of that medical school.

Eric Kandel also sought advice from Grundfest. As an undergraduate at Harvard, Kandel came to know and admire several psychoanalysts. In his third year of medical school at New York University,

Figure 1. The New York Neurological Institute.

in the nursery for newborn infants. This traffic has been lessened by the method of not bathing and nonoiling the newborn. The time spent each day by the nurse hovering over the baby is greatly lessened. In the average charity hospital or county hospital, where the infant population varies from eighteen to fifty a day, suggestions 1, 2 and 3 are hardly practical, but hospital delivery and discharge within two to three days would seem highly desirable. This practice has been carried out in certain hospitals with excellent results. Of course, the ultimate of dispersion is home delivery, but I cannot picture an obstetrician with a trailer hitched to his car carrying delivery equipment going out to deliver a baby in the home. Dr. Spivek deserves a medal from the Section on Pediatrics for suggesting such a practical method.

Dr. Maurice L. Blatt, Chicago, Ill.: Dr. Spivek's subject is most timely in view of the fact that epidemics of sepsis have invaded the nurseries for the newborn of institutions from New York to the Pacific. San Francisco apparently has been fortunate in not having had a serious epidemic in any of its nurseries. The problem as presented by Dr. Spivek is a result of our type of civilization. The assembling of large groups of people in a single hospital endangers each by intimate contact with carriers or individuals with disease. To disassemble such a civilization is difficult and it is this which Dr. Spivek suggests. I do not believe that advance can be made along the lines so suggested. I believe it is up to us as scientists, as pediatricians, as research workers, to determine a more effective method of prevention of sepsis in the newborn in nurseries in large communities. Nurseries will continue to be large because in no other way can we finance the care of the large number of indigent and semi-indigent patients thrust upon us.

Dr. M. K. Wylder, Albuquerque, N. M.: I come from a state where 40 per cent of the population live long distances from a doctor, where about 30 per cent of our women are delivered not even by a physician, and over 50 per cent of the women in our state are delivered in the homes. I see a great many of these babies that are delivered by a dirty Mexican midwife who washes her hands after the delivery instead of before it. However, in cases in which delivery is done on a sheep pelt laid under the woman in the bed, impetigo and septic infections do not result. While I don't think any of us want to go back to home delivery, impetigo and septic infections in the newborn in the home delivery aren't seen by us. This, then, is purely a problem for the hospital. As Dr. Arthur Abt told the round table at Delmonte the other day, if the hospitals don't solve these problems, the deliveries will go back to the home again. I believe the hospitals will solve them.

Dr. Irving J. Wolman, Philadelphia: I think all pediatricians who are familiar with the problems of the nursery would agree with Dr. Spivek whole-heartedly. One big problem, of course, lies in the etiology of these gastrointestinal infections which are now making their appearance all over the country. We younger men wonder whether these epidemics are new things or whether they have been going on for years without being reported in the literature. Most hospital routines are based on the assumption that the infective agents spread by direct contact and fomites. A most important but hitherto neglected approach lies in the prevention of air borne infections by the newer methods of air hygiene. It is questionable whether restricting the number of infants in any one nursery to two or four is sufficient. It would be better to isolate each individually. There are some new developments in the field of air conditioned cubicles within which infants are completely protected and isolated from the outside world. Dr. Chapple of Philadelphia presented a description of one such chamber before the American Pediatric Society last month. His rigid isolation technic includes the breathing of air from outdoors where the bacterial count is very low. The addition of atmospheric sterilization to the nursery armamentarium may prove of service in the control of nursery epidemics.

Dr. Mandel S. Spivek, Chicago: This paper is a theoretical one, a paper which was gotten up at a desk. It has nothing to do with research or with a reading of literature. It is my own appraisal of what goes on in the nursery. Dr. Wolff has

a wonderful setup at the hospital with which he is associated, in which one of the most important things is the esprit de corps. That is, every one is aware of the technic, every one is conscious of it. There are very few hospitals that have that mutual understanding. Under those conditions, the spread of infection is difficult. Whether we can go back to home delivery or not, I do not know. It took a lot of educating and a lot of pushing to get the mothers from the home into the hospital. It will take the same education and the same push to get them back to their homes for delivery. It is merely a matter of education. It should not be difficult; I don't see any reason why we can't. Dr. Wolman spoke of air borne infection. We really do not know whether it is air borne or not or whether it is a virus disease or not. Various organisms have been reported, and in a thorough search for the virus in the Chicago epidemic nothing was found.

SODIUM DIPHENYL HYDANTOINATE IN THE TREATMENT OF CONVULSIVE DISORDERS

H. HOUSTON MERRITT, M.D.
AND
TRACY J. PUTNAM, M.D.
BOSTON

Good results in the treatment of patients with convulsive seizures have been obtained by a variety of methods, such as medical treatment with bromides or barbituric acid compounds, the ketogenic diet, restriction of fluids, and the surgical excision of scars or irritable cortical foci. In spite of these various therapeutic means there are a great number of patients who are not relieved of their attacks or are helped only temporarily by treatment. The fact that treatment by anticonvulsant drugs is at present the most widely used and on the whole the most effective method of therapy suggested to us the possibility that a direct and systematic experimental search might reveal more potent and less sedative compounds. With this idea in mind we devised an experimental procedure[1] which would produce convulsive seizures in animals at a constant threshold and which would allow for a qualitative and roughly quantitative determination of the relative effectiveness of various drugs. By this method a large number of chemicals and drugs, selected on theoretic grounds, were studied, and it was found[2] that several heretofore unused chemicals were as effective in protecting animals from the electrically induced convulsions as the drugs commonly used in the treatment of patients with convulsions (bromides, phenobarbital and the like) or even more effective.

Chemical structure of sodium diphenyl hydantoinate.

Under the auspices of the Harvard Epilepsy Commission.
This study was aided by a grant from Parke, Davis & Co., Detroit.
From the Neurological Unit, Boston City Hospital, and the Department of Neurology, Harvard Medical School.
Read before the Section on Nervous and Mental Diseases at the Eighty-Ninth Annual Session of the American Medical Association, San Francisco, June 17, 1938.

1. Putnam, T. J., and Merritt, H. H.: Experimental Determination of the Anticonvulsant Properties of Some Phenyl Derivatives, Science 85: 525-526 (May 28) 1937.
2. Merritt, H. H.; Putnam, T. J., and Schwab, D. H.: A New Series of Anticonvulsant Drugs Tested by Experiments on Animals, Arch. Neurol. & Psychiat. 39: 1003 (May) 1938.

Figure 2. The first page of JAMA article describing safety and efficacy of Dilantin in treating epilepsy. This paper (JAMA 1938; 1068) was designated a "Landmark Article" (JAMA 1984; Feb 24;251(8):1062-7).

Figure 3. Derek E. Denny-Brown
(1901–1981)

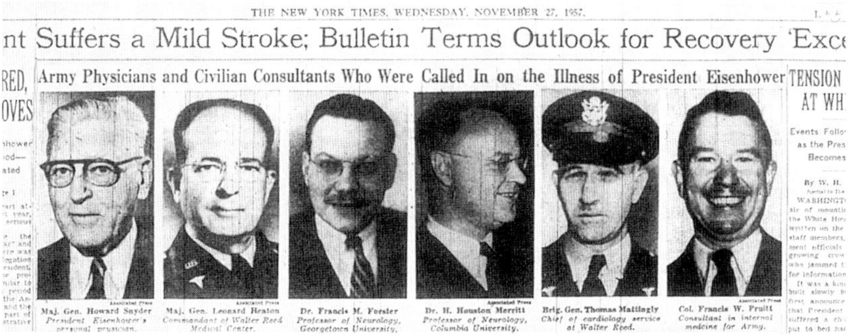

THE NEW YORK TIMES, WEDNESDAY, NOVEMBER 27, 1957.

nt Suffers a Mild Stroke; Bulletin Terms Outlook for Recovery 'Exce

Army Physicians and Civilian Consultants Who Were Called In on the Illness of President Eisenhower

Maj. Gen. Howard Snyder *President Eisenhower's personal physician.*

Maj. Gen. Leonard Heaton *Commandant of Walter Reed Medical Center.*

Dr. Francis M. Forster *Professor of Neurology, Georgetown University.*

Dr. H. Houston Merritt *Professor of Neurology, Columbia University.*

Brig. Gen. Thomas Mattingly *Chief of cardiology service at Walter Reed.*

Col. Francis W. Pruitt *Consultant in internal medicine for Army.*

Figure 4. Merritt was one of several distinguished physicians called to examine
President Dwight D. Eisenhower who had an ischemic stroke.

Figure 5. *Left:* Frederick Gibbs. *Right:* William Lennox.

Figure 6. Portrait of Edward S. Harkness.

Figure 7. Henry Moses, President, Board of Trustees, Montefiore Hospital, recruited Merritt to New York City in 1945.

Figure 8. A tribute to H. H. Merritt.

Figure 9. *Left to right:* William Lennox, Emma Gibbs, and Frederick Gibbs. From Lombroso C., *Epilepsia* 1988; 29 suppl 2: S5-14.

Figure 10. Bust of William G. Lennox; copies at Lennox Epilepsy Centers in Japan, Belgium, and Boston.

Figure 11. Otto Marburg, one of the German Jewish refugees saved from the Nazi government by Tracy Putnam.

Figure 12. Merritt's Last Case Conference, 1967. The pseudo-patient (a neurosurgical resident) had every known adverse effect of Dilantin. Courtesy of Dr. Roger N. Rosenberg.

Figure 13. H. Houston Merritt. He listened attentively and read the patient's history carefully.

Figure 14. Merritt loved to play poker. He refused to end a game until he won. Note that the loot includes a wristwatch.

Figure 15. Grass Foundation trustees. *Left to right:* Albert Grass, Frederic Gibbs, Ellen Grass, Robert Morison, and Erna Gibbs (not a Trustee). From Zottoli SJ. *Biol Bull* 201: 218–226. (October 2001)

A MODEL BOY IN TOYLAND

Figure 16. Another tribute to Merritt.

A NEW SERIES OF ANTICONVULSANT DRUGS TESTED BY EXPERIMENTS ON ANIMALS

H. HOUSTON MERRITT, M.D.

AND

TRACY J. PUTNAM, M.D.

WITH THE TECHNICAL ASSISTANCE OF DOROTHY M. SCHWAB, A.B.

BOSTON

A method for the determination of the convulsive threshold by means of graded electrical stimulation has been described by Spiegel [1] and has been employed for comparing the action of certain drugs. The apparatus used in this investigation represents a simplification of that devised by Spiegel and embodies also the arrangement of electrodes employed by Krasnogorski.[2] A description and diagram have been published.[3]

The point of departure for the investigation was the fact that although phenobarbital is one of the most efficient anticonvulsant drugs in common use, other barbiturates are comparatively ineffective, a fact that is often observed clinically and is strikingly demonstrated by the apparatus employed. For this reason and from certain theoretical considerations,[3] a search was made among phenyl derivatives of the general type of phenobarbital, including phenyl, cresyl and tolyl sulfonates, benzoates, ketones and esters, with such radicals as carbamic, barbituric and malic acid, and hydantoin.

METHOD

Cats were used as experimental animals. The threshold for convulsions was determined by applying increasing intensities of an interrupted current from an electrode placed in the mouth to one on the moistened hair of the occiput. The current was measured by a 0-50 milliammeter in the circuit and was drawn from

This work was aided by a grant from Parke, Davis & Co., Detroit.

Read at the Sixty-Third Annual Meeting of the American Neurological Association, Atlantic City, N. J., June 5, 1937.

From the Neurological Unit, the Boston City Hospital, and the Department of Neurology, Harvard University Medical School.

1. Spiegel, E. A.: Quantitative Determination of the Convulsive Reactivity by Electrical Stimulation of the Brain with the Skull Intact, J. Lab. & Clin. Med. 22:1274 (Sept.) 1937.

2. Krasnogorski, N. I.: Tr. Internat. Physiol. Cong., 1935, p. 213.

3. Putnam, T. J., and Merritt, H. H.: Experimental Determination of the Anticonvulsant Properties of Some Phenyl Derivatives, Science 85:525 (May 28) 1937.

Figure 17. New series of anticonvulsant drugs (*Arch Neurol* 1938;39:1003–15).

TRACY JACKSON PUTNAM, M. D.
NEUROLOGICAL SURGERY
450 NORTH BEDFORD DRIVE
BEVERLY HILLS, CALIFORNIA
CRestview 4-5304

May 16, 1961

Mr. Meyer Friedman
411 North Palm Drive
Beverly Hills, California

Dear Meyer:

You have asked me to prepare for you an outline of
my training and experience in research, invention and
administration. I am enclosing herewith my curriculum vitae
and list of publications, in the form usually used by medical
investigators and teachers, but perhaps your needs would be
better met by a narrative summary.

I graduated from Harvard College in 1916. This
was a year later than the rest of my class, because I spent
a year with the French Army, obtaining a Croix de Guerre.
I entered the Harvard Medical School at once. While there,
I obtained my first patent, a simple device for shaking blood
count pipettes. I also did my first original scientific work,
which included describing for the first time a tiny structure,
visible to the naked eye, in the human brain. I graduated
first in my class.

The next year, I took an assistant residency in
pathology at the Johns Hopkins Hospital, then an internship
in surgery, then a residency in neurological surgery. In 1925,
I went to Amsterdam to study neuropathology, and was assigned
an anatomical problem which involved the presentation of
information in these dimensions. I also studied psychology and
psychiatry there.

On returning to Boston the next year, I was put in
charge of the surgical research laboratory. The problems
which I chose were chiefly in the fields of chemistry and
endocrinology. In the course of these, I had to set up a
miniature production line for the preparation of a new hormone.

In 1929, I transferred my activities from the
Department of Surgery to the Department of Neurology, where
I was promised a position as neurological surgeon. While
waiting for proper surgical facilities to be constructed,

Figure 18a. Putnam's autobiographical letter p. 1. Courtesy of Darling Biomedical Library, University of California, Los Angeles.

I accepted a research assignment in the field of multiple
sclerosis, a subject I have pursued ever since. My first
work was to read virtually everything which had been written
about it, over a thousand books and articles, many of them in
foreign languages. I then initiated a series of experiments
and observations, and did practically nothing else for the
next two years.

In 1931, I began work as an independent neurological
surgeon, and developed an effective operation for the very
distressing diseases characterized by involuntary movements
(athetosis, dystonia, chorea). I also carried out some studies
on the circulation of the brain. This was in the depths of
the depression, and money for research work was scarce indeed.
In order to measure the blood flow in a monkey's brain, I
improvised an apparatus out of a fuel pump, a windshield
wiper and an erector set. In these years also, I succeeded in
producing multiple sclerosis in animals for the first time.

In 1933, I accepted an Assistant Professorship in
Neurological Surgery and a position at the Massachusetts
General Hospital. Here I designed and constructed a new and
useful endoscopic instrument for operating on babies suffering
from hydrocephalus (water on the brain). This instrument solved
the problem of maintaining adequate vision through a fluid
rendered cloudy by blood.

In 1934 the position of full Professor of Neurology
in charge of the clinic and extensive laboratories at the
Boston City Hospital was offered to me, and at some financial
sacrifice I accepted it. I immediately set up the first
routine clinical laboratories for electroencephalography and
for electromyography in the world, where any doctor could send
a patient and have a report. Up to this time, both of these
techniques had been laboratory toys.

I also looked into the possibility of finding a
drug which would control epilepsy without being sedative.
From a consideration of various types of data, I concluded
that one might be found, predicted the approximate formula,
and devised a simple method of measuring anticonvulsant
activity. The result of this was the drug now known as
dilantin, which is still the most widely useful of its series,
and my test has become a standard one in pharmacology.

The work on multiple sclerosis was continued, and
led to a fairly complete picture of the mechanism of the

Figure 18b. Putnam's autobiographical letter, p. 2.

disease. I was able at that time to predict the use of
most of the measures of treatment which have since been found
useful.

I also initiated the operative treatment of
paralysis agitans, and devised a number of laboratory gadgets,
stains, modifications of photographic procedures, etc. I
described the pathological and clinical effects of anoxia on
the brain.

In 1939, I was offered the chairs of Neurology and
Neurosurgery at Columbia University and the control of the
Neurological Institute. This was a much larger service,
the laboratories were marvelous, and the budget was about
six times as large as I had had in Boston. With some misgivings,
I accepted. The work involved was staggering. The clinical
services were archaic, and plagued with jealousies; the
laboratories had long been inactive; and the war in Europe broke
out just as I arrived. I was appointed to the National Research
Council, to the Health Insurance Plan, to the Board of
Neurology and Psychiatry and that of Neurosurgery, and became
the neurological consultant to a group from Columbia investigating
the effects of altitude and anoxia in one of the early pressure
chambers. All of this cut down my opportunities for doing
personal research work.

I also became Vice President of the Committee for
Resettlement of Displaced Foreign Physicians.

From 1940 to 1946 the Neurological Institute
developed from being the laughing-stock of the profession to
the foremost establishment of its kind in the world, and my
department became the most productive one in the medical
school. I established a department of electroencephalography
there also and undertook a study of frequency analysis of
brain wave patterns, which are composed of frequencies from
2 to 30 per second. In the course of this, I constructed a
series of electromechanical tranducers to cover this frequency
range. One of them served as an automatic indicator of the
onset of anoxia in the pressure chamber, but it was cumbersome
and the Council on Aeronautics was not interested. Nor were
they interested in the possibility that dilantin might mitigate
the symptoms of anoxia, which I suggested. This was later
demonstrated by others.

Figure 18c. Putnam's autobiographical letter, p. 3.

We also developed the first of another series of anticonvulsant drugs, namely glutamic acid. This and its derivatives are still being investigated by other workers.

Work on multiple sclerosis continued, and as soon as the anticoagulant drugs were developed, they were successfully used in treatment.

We carried out two extensive investigations for the National Research Council. One was a study of the possibilities of bacteriological warfare. The other was a thorough study with accurate special instruments and tests, of the various effects of concussion and contusion of the brain.

In one of my reports to the Committee on Anoxia, about 1942, I suggested the use of a pressurized ejection capsule provided with a parachute for high altitude flight.

In 1943, Mr. Charles Cooper was elected President of the Presbyterian Hospital. He at once began a campaign to reduce the Medical School to an appendage of the hospital, and made a few menacing gestures in my direction. In 1945, he (presumably) sent me word through Dr. Robert Loeb, who was then being groomed for Professor of Medicine, that I should get rid of all the Jews in my department or resign. I replied that I had no intention of doing either, and that he could not force a professor to resign; but I was wrong about the latter. The faculty meekly fell in line with Cooper's instructions, despite a few scattered protests, and I saw that I should have to leave. This was a source of much distress to me, for with the end of the war I visualized enormous possibilities for scientific advancement. They have not been realized, and the neurological department at Columbia has lapsed back into the state of anxious inactivity which I found there in 1939.

The final break came in 1946, and I moved to California to go into private practice. Since then, I have devised a new treatment for recalcitrant cases of epilepsy, a new type of stereotaxic guide for operations on the brain, and a new and useful operation, replacing prefrontal lobotomy for the relief of intractible pain and anxiety, without significant impairment of personality.

In 1946, I founded the National Multiple Sclerosis Society. I have never been able to get any funds for research from it, however. In the same year, I also founded what has since become the National Epilepsy Society.

Figure 18d. Putnam's autobiographical letter, p. 4.

From 1946 to the present, I have become more and more interested in the commercial and possible military applications of some of the techniques which I learned while pursuing medical research. About 1952, I obtained a patent for a new type of elevator with a variable-ratio fluid drive, provided with a pressure tank to store up the energy produced by descent of a load, to aid in lifting the next. I also patented an electronic device to teach dogs not to bark.

In 1952, I became involved in a partnership in a machine shop with a brilliant but unstable engineer, who soon became very ill, dying early in 1954. This was a financial catastrophe for me, from which I am only just recovering. It led me to a study of production methods, however, and I outlined designs for a number of new tools. They include: a drop hammer with the characteristics of a hydraulic press; a pistonless internal combustion pump; a principle which can be applied to such tools as a pickaxe or a plow, enormously increasing the effectiveness of either by means of an explosive force; a hydraulic powered garbage disposal unit; a non-hydraulic high pressure press; a portable irrigating device, which will water several acres at a time; and others.

About this time also, I suggested the use of a magnetostrictive drill in dentistry, and the use of magnetic dentures, both of which have since been introduced by others.

In 1954, through a friend, I became interested in the scientific grading of fur, and have applied for a patent for a device which will facilitate matching the color of furs. I also stumbled on another principle which will greatly enhance the beauty of fur on display. I think I know how to measure the sheen and the density of fur. So far, these ideas remain unproductive, for lack of capital.

In 1958, with Dr. John Button, I inserted electrodes into the visual cortex of a blind girl, and by stimulating them, was able to give her the sensation of flashes of light. I constructed an oscillator controlled by a photocell, by means of which she could find her way around a room. I have in mind a number of further steps along these lines, and have filed an application for a patent. We were unable to secure research funds, and the investigation is at a standstill.

I have shared in the ownership and development of a little machine for counting pills, suitable for use in retail drug stores.

Figure 18e. Putnam's autobiographical letter, p. 5.

Mr. Meyer Friedman
5-16-61
-page 6-

You already have a description of the "Supersnooper", and a list of projects founded on the "Inventions Wanted" book of the Armed Services, so this brings us to the present.

In summary, I think I may say that I have a fair grasp of classical physics, chemistry, electronics, mechanics, physiology and psychology, which has been gained through individual practical experience and reading original sources, rather than through courses. I have perhaps done best when working in fields where several disciplines overlap. When I was in an administrative position, I had the reputation of making funds go further, and getting gifted individuals to work more productively than many others in similar positions. I have come to a time of life when the practice of medicine is no longer the adventure it used to be, and I should enjoy developing some new ideas.

Always yours,

Tracy J. Putnam, M.D.

TJP:dd

Figure 18f. Putnam's autobiographical letter, p. 6.

Figure 19. Tracy J. Putnam (1894–1975), ambulance driver in France during World War I.

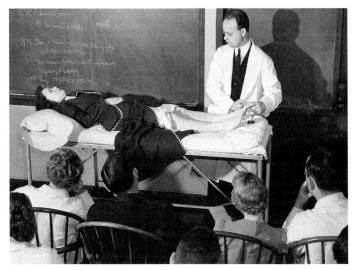

Figure 20. Saturday morning at Boston City Hospital. Courtesy of Drs. John Steiner and Charles D. Aring, University of Cincinnati.

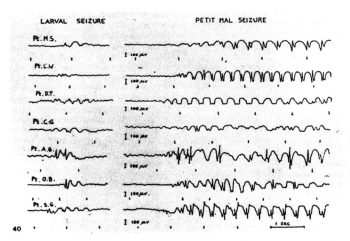

Figure 21. Early recording of EEG in petit mal epilepsy. From Gibbs F, Davis H, Lennox WG. *Arch Neurol Psychiatry* 1941;34:1133–48

Figure 22. NINY Neurology and Neurosurgery house staff, 1950. Houston Merritt and J. Lawrence Pool are seated in center of front row. Courtesy of Drs. T. A. Pedley and S. DiMauro.

Figure 23. Lucy G. Moses was a benefactor of the Department of Neurology at the Neurological Institute. When she visited, the staff greeted her royally. Courtesy of Drs. T. A. Pedley and S. DiMauro.

Figure 24. Frederick Tilney (1876–1938).

Figure 25. Houston Merritt's childhood home, Wilmington, NC.

Kandel began to hear about Grundfest and sought advice. He wanted to know where in the brain were the ego, id, and superego or "where in the elaborate folds of the human brain these psychic agencies might live."[13] Grundfest responded that the question could not be answered and that, to understand the mind, they needed to study the brain "one cell at a time."

Later, Kandel worked in Grundfest's laboratory with Purpura, and Grundfest arranged for Kandel to work in the NIH laboratory of Wade Marshall, where Kandel met Alden Spencer and they set off trying to learn about learning and memory, an effort that ultimately led to Kandel's research with a simple organism, the sea slug *Aplysia*. By that time, Spencer had died of amyotrophic lateral sclerosis, but the stars were otherwise certainly in an auspicious constellation for these budding investigators and a future Nobelist.

Kandel is not only a remarkable neuroscientist but a fantastic writer. With James Schwartz and Thomas Jessell, he has edited and written the standard textbook on neuroscience for medical students and graduate students. Kandel's *In Search of Memory*[13a] was not only a best seller but also the choice of other scientists for the 2007 National Academies of Science Best Book Award. The book is a triple value—standard autobiography, explication of his work on the cellular basis of memory, and a brief but informative history of neuroscience, all written with verve and clarity for the general reader.[14]

Frederick A. Mettler (1907–1984)

Frederick A. Mettler was the fourth of Tracy Putnam's basic science recruits. He was the only one who was not elected to the National Academy of Sciences, but he had a major impact in what is now called "functional neurosurgery." More specifically, he was one of the earliest neuroanatomists who aimed to control the involuntary movements that may arise from damage to specific parts of the brain. He analyzed the nerve pathways involved in the movements and then designed operations that would interrupt these nerve tracts and, thus, ameliorate the movements. His work led to the current practice of deep brain stimulation for the amelioration of the symptoms of Parkinson disease.

Mettler wrote on many subjects, including several biographies and obituaries for the American Neurological Association and the *Journal of Comparative Neurology*; but his own death went unnoticed, with not a single recorded biography or appreciation.

Mettler's family originated in Switzerland and was presumably prosperous enough for him to attend a private school, Blair Academy, in New Jersey. He graduated from Clark University in 1929 and gained a PhD in anatomy from Cornell in 1933. He served for one year at St. Louis University and then went to the University of Georgia as an assistant professor of Anatomy, while also being a medical student. He received his MD degree in 1937. Teaching all the while, he was promoted to full professor of Anatomy in 1938 and then recruited to Columbia in 1940; presumably, Putnam had some role in that.

Throughout this period, Mettler published prolifically on the fiber tracts in the brain of macaque monkeys, and he started work on a textbook of neuroanatomy, which was published soon after the move to Columbia. That book was soon followed by a 1947 book, which combined Mettler's neuroanatomy with clinical chapters written by Merritt and Putnam.[15]

Mettler had become well known by this time and began to attract research fellows, among them Malcolm Carpenter, John Whittier, Robert Heath, and this author. Each of them attained recognition, starting with their research in Mettler's laboratory late in the 1940s.

Whittier and Carpenter made lesions in monkeys to induce involuntary movements and then made remedial incisions to relieve the movements. Carpenter followed Mettler and other Columbia neuroanatomists to become responsible for teaching and also writing his own highly popular textbook. Whittier moved to the Creedmore Psychiatric Institute in the borough of Queens, New York, and helped established the Huntington Disease Foundation; that disease is characterized by involuntary movements that are called "chorea." In 1983, the disease was the first to have the gene mapped by the then new methods of molecular genetics, but it took another decade to identify the mutated gene product.

Heath became one of the more controversial figures in modern neurology and psychiatry. He made innovative educational changes as chair of both Neurology and Psychiatry at Tulane University in New Orleans, and he was a pioneer in biological psychiatry, implanting electrodes into the human brain and studying blood constituents. However, he erroneously claimed to have found a protein in the blood that, if injected into the blood of another person, could cause schizophrenia. That discovery could not be repeated by other investigators and raised ethical questions about human experimentation. Heath was also said to have collaborated with the CIA in experiments on mind control.[16,17]

Mettler later became interested in psychosurgery. He recognized the absence of specific information about the anatomical structures that were affected by the operation and the difficulty of interpreting results. He therefore led the Greystone Project, working with patients at a state mental hospital in New Jersey and with J. Lawrence Pool, head of neurosurgery at Columbia. They excised limited pieces of cortex from different parts of the frontal lobes in an operation they called "topectomy." The operation was at once more limited anatomically than lobotomy, and therefore more precise, and less likely to have undesirable psychiatric effects. Mettler had become an authority in yet another field, but lobotomy fell into disfavor and so did topectomy.

Mettler retired from Columbia in 1975 to spend the rest of his days at his farm, Pippin Hill. He developed pulmonary fibrosis, which led to his death in 1984.

Mettler's first wife was Cecilia Mettler, a medical historian. She died after the birth of their first child, also named Cecilia. Fred completed his wife's unfinished book on the history of medicine. His daughter ultimatly became a professor of laboratory medicine at the University of Cincinnati School of Medicine. Mettler remarried and had a son, Frederick, Jr. But his new wife was not around much and Fred, Sr., brought up the two children by himself. Fred, Jr., also had a distinguished career as a professor of radiology at the University of New Mexico and a world authority on the biomedical effects of radiation at Chernobyl and in diagnostic radiology.

Mettler, the anatomist, had a wry sense of humor. He was the official neuroanatomist for the Bronx Zoo and once did an autopsy on a tiger that died there. One day, he summoned a research fellow to his office and told the young man that a hyrax had died; the brain was to be retrieved and returned to Mettler. The young man put the tools together and, on the way out, asked Mettler, "What is a hyrax?" Mettler just pushed him out the door.

The brain was indeed retrieved at the zoo. The amateur patholo-gist found out that the animal was an interesting anomaly in evolu-tion. It was the size of a raccoon, had teeth like a rodent, but had hoofs like a cow (or a pig).

The brain was put in formalin and carried by hand back from the Bronx to 168th Street and delivered to Mettler, who was delighted and greeted the fellow with a question, "So, tell me now, just what is a hyrax?" [18]

Notes

1. Anonymous. Medicine. Bread-&-butter brains. Time, October 16, 1939, http://www.time.com/time/searchresults/

1a. Tilney F, Riley, HA. Form and function of the nervous system: an introduc-tion to the study of nervous diseases, 2nd ed. New York: Hoeber, 1923.

2. Huxley A. Kenneth Stewart Cole (July 10, 1900–April 18, 1984). Biographi-cal memoirs, vol 70. Washington DC: National Academies Press, 1996, pp 25–45.

3. Kabat EA. Getting started 50 years ago—experiences, perspectives, and problems of the first 21 years. Annu Rev Immunol 1983;1:1–32.

3a. Ibid.

4. Kabat EA, Hanger FM, Moore DH, Landow H. The relation of cephalin flocculation and colloidal gold reactions to the serum proteins. J Clin In-vest 1943;22;563–568.

5. Kabat EA. Before and after. Annu Rev Immunol 1988;6:1–24.

5a. Kabat EA. Before and after. Annu Rev Immunol 1988;6:1–24.

6. Mage RG, Feizi T. Elvin A. Kabat (1914–2000). Biographical memoirs, vol 85. Washington DC: National Academies Press, 2004, pp 3–27.

7. Long T. Elvin A. Kabat, at 85: researcher received national science medal [obituary]. Boston Globe, June 21, 2000.

8. Marcus DM, Schlossman SF. In memoriam. Elvin Abraham Kabat (Septem-ber 1, 1914–June 16, 2000). J Immunol 2001;166(6):353–356.

9. Saxon W. Elvin Kabat, 85, microbiologist known for work in immunology. New York Times, June 22, 2000.

10. Paul WE, Mage RG. Elvin A. Kabat (1914–2000). Nature 2000;407:316.

11. Khiss P. Columbia professor gets McCarthy contempt threat. New York Times, November 26, 1953.

12. Reuben JP. Harry Grundfest (January 10, 1904–October 10, 1983). Bio-graphical memoirs, vol 66. Washington DC: National Academies Press, 1995, p 150.

13. Kandel ER. In search of memory. The emergence of a new science of the mind. New York: W. W. Norton, 2006, p 55.

13a. Ibid.

14. Rowland LP. A Nobel laureate offers both personal history and a scientific tour de force on memory. Neurology Today 2006;6(8):34–35.
15. Merritt HH, Mettler FA, Putnam TJ. Fundamentals of clinical neurology. Philadelphia: Blakiston, 1947.
16. Mohr CL, Gordon JE. Tulane: the emergence of a modern university, 1945–1980. Baton Rouge: Louisiana State University Press, 2001, pp 120–123.
17. Valenstein E. Brain control; a critical examination of brain stimulation and psychosurgery. New York: John Wiley, 1973, pp 164–168.
18. Personal recollection of author; the author was "the young man."

8

Putnam Gets the Boot:
How It Was Done

When the Presbyterian Hospital in the City of New York joined with the Columbia University College of Physicians & Surgeons, they adopted guidelines that have remained in place ever since. Chairs of the departments are university officers, and they are the ones who nominate new members of the faculty. If a clinical department is involved, the person has two titles, one at the university and the other in the hospital. For instance, an assistant professor of neurology in the medical school is an assistant attending neurologist in the hospital.

In the medical school, Tracy Putnam's title was professor of neurology. At Presbyterian Hospital, he was director of two separate services, Neurology and Neurosurgery. He held these positions with "tenure," which ordinarily means that he could not be forced to relinquish his positions without cause. Tenure was designed to protect the right of university professors to speak and write without fear of being fired for expressing unpopular views. Tenure is granted by the university, not the hospital. It is a promise to pay a salary if other sources are lost but has never been tested at Columbia as far as the author knows.

If a clinical chair is deemed to be wanting as an administrator, he or she may be asked to resign, which has happened several times since the Putnam affair. However, when Tracy Putnam was attacked, as we will see, Dean Rappleye reiterated several times his belief that the Putnam problem had originated and was "entirely a hospital matter." Nevertheless, university committees had to agree, and so they did.

When Tracy Putnam became director of the Neurological Institute in 1939, the hospital president was Dean Sage, Jr. (1875–1943). "Dean" was his first name, not a title. Putnam wrote that his relationship with Sage was always smooth and that he had no problems until Sage died in 1943. Sage came from a family with wealth gained from the timber industry. Among his forebears were several New York State senators. As an undergraduate at Yale, Sage was a member of the exclusive secret society Skull and Bones. A website[1] provides his biography, starting with the following statement: "The Order has dominated the board of trustees [of Presbyterian Hospital] from at least 1922 to 1957." Active as a lawyer and director of the Sage Land and Improvement Company, he became president of Presbyterian Hospital in 1922 and held that position until he died in 1943. He was also director of the Commonwealth Fund and the Josiah Macy, Jr. Foundation, both heavily involved in medical education.

According to Putnam, his problems commenced with the 1943 death of Sage and the arrival of Charles Proctor Cooper (1884–1966) as president of Presbyterian Hospital. Trained as an engineer at New Hampshire State College, he joined the New York Telephone Company in 1908, one year after graduation. As phone service expanded, so did his responsibilities, and in 1926 he was vice president of AT&T. By 1946, he was executive vice president. He retired from the phone company in 1948. His association with the medical center began in 1930 as a trustee of the Neurological Institute, then the board of Presbyterian Hospital in 1938 and president of the hospital from 1943 to 1957. He was also an executive of an insurance company and a bank.

What happened to Putnam is recorded in the files of both the hospital and the dean's office. Dean Willard C. Rappleye was busy in the fall of 1944, only five years after Tracy Putnam had been hired.

On Friday, September 15 Dean Rappleye conferred with President Cooper and Vice President John Parke, the hospital leaders. The next Monday Rappleye met with Putnam. These meetings were repeated several times in the next two weeks. On October 3, Byron Stookey and Henry A. Riley came separately, and then the dean met with the Medical Board of the Neurological Institute. There followed several meetings with Putnam and one more with Cooper and Parke on October 20. Then, on October 30 and 31, Stookey and Riley returned again separately. In November there were more meetings with Parke. Edwin Zabriskie, a much respected clinician, came several times on "unscheduled appointments." Putnam came for the last time on November 20, and on December 1 "Dr. Zabriskie assumed administrative responsibilities for which he was paid a salary."[2]

The hospital version of those events was recorded separately, but it was written three years later and in the following chronology:

September 15, 1944: Charles P. Cooper, president of the Presbyterian Hospital, and Mr. John S. Parke, executive vice president, presented to the dean verbally their decision that the administration of the Neurological and Neurological Surgery services in the Neurological Institute was "unsatisfactory." Subsequent to the discussion, the dean talked with Dr. Putnam, who said there was "no ground for this dissatisfaction." Cooper referred the matter to the Medical Board.

October 26, 1945 (one year later): "At a special meeting of the Executive Committee of the Medical Board, the Chairman reported that Presbyterian Hospital is of the opinion that the administration of the Neurological and Neurosurgical Service is unsatisfactory and that the present Director, Dr. Tracy Putnam, should be requested to relinquish his responsibilities as the Executive of those Services. After full discussion, the Executive Committee offered no objection to this request made by the Presbyterian Hospital."[3] The implications of this statement are clear. Service directors went along with the hospital administrators' antipathy to Putnam. He was never to recover.

November 1945: A special committee of the Medical Board was appointed, with Albert Lamb as chair;[3a] the minutes of that meeting were recorded in a 1947 meeting of the Committee on Administration, a joint committee of the medical school and hospital, and included the events from the start in 1944.[4] The committee interviewed staff members of the Neurological Institute and concluded there was "a good deal of dissatisfaction with the administration of professional services in the Institute from a purely administrative point of view."[4a] It was the consensus of practically all of those consulted, as well as committee members, that the combined services were too large for active administration by a single individual, especially one who was also engaged in private practice. The committee reported to Cooper on February 13, 1946, its recommendation that the two services (neurological surgery and medical neurology) be separated on the basis that "it is as illogical for one man to cover adequately both services as it would be for one man to cover both general Medical and Surgical Services in the Presbyterian Hospital."[5]

February 13, 1946: The special committee reported to Cooper, recommending separation of the neurology and neurosurgical services. They concluded that Putnam should continue as head of Neurology, but he responded that if the services were to be separated, he preferred to lead Neurosurgery. After discussions between Putnam and staff members, Cooper, and the dean without a final settlement of the problem, Putnam requested a leave of absence beginning on July 1, 1946.

April 4, 1946: This was reported to Cooper, who verbally agreed to the proposal.

December 16, 1946: Another "informal" conference was attended by three service directors, Drs. George Cahill (Urology), Walter Palmer (Medicine), and Rustin McIntosh (Pediatrics), with Dean Rappleye and Mr. Parke, to verify the plan to separate the services. The dean spoke to Putnam the next day and thought Putnam would return after his leave of absence, as director of Neurological Surgery. Putnam had moved to Los Angeles.

December 18, 1946: Dr. Edwin Zabriskie also phoned Putnam and thought that Putnam, fully aware of the plan, would return to duty on July 1, 1947. Zabriskie also understood that Putnam was writing a letter to the dean to confirm their conversation.[5a] These matters were presented to the Committee on Administration to ratify the separation of services and the appointment of Putnam in Neurological Surgery, effective July 1, 1947. The committee also recommended the appointment of a new head of Neurology by that date, assuming that Putnam would be notified in February 1947 when he returned from Los Angeles.

February 1947: The situation was festering. Putnam had not resigned, had requested a leave of absence, and had moved to Los Angeles. Rappleye sent the minutes of that meeting to Frank D. Fackenthal, acting president of Columbia University, with a covering letter in which he wrote the following:

There was nothing irregular about what we did because we have
to begin making some arrangements for July 1st if we are to have
someone heading our Department of Neurology. Dr. Putnam,
who has been in town on several occasions, has never come to
talk to us about our appointments or budget and we have been
operating somewhat in a vacuum. Nevertheless, we have been
giving every consideration to Dr. Putnam. There is no one here
who has the slightest doubt about the necessity of separating the
services of neurology and neurological surgery since the Hospital authorities and the Medical Board have unanimously agreed
that they should be separated in the Neurological Institute. Dr.
Putnam understood this before he left."[5b]

March 5, 1947: Rappleye met with Putnam in a "long and friendly
conference."[6] Putnam had concluded that he had to resign from
his hospital and university positions as of June 30, 1947, when
his leave of absence would end. He would submit the letter of
resignation within three weeks of his return to California. Rappleye agreed to the transfer of research funds, apparatus, and
books.

April 13, 1947: Putnam wrote the following to Rappleye from Los
Angeles:

I intend to resign as of June 1. . . . I know that you will not
think that I am doing this out of pique, but because of a deep
conviction that no matter what compromise I may be willing to
make, the Administration will not be satisfied, and will continue
to badger me with cause, as in the past. You have been witness to
the deeds and words on which this conviction is based.[7]

Rappleye responded and concluded, "Certainly you go with my heart-
iest good wishes and warmest personal regards."[8]

June 30, 1947: Putnam submitted an eight-page letter of resignation, a temperate and detailed defense of his administration.[9]
Sending the letter to hundreds of people, he wrote that he had
been hired to make the Neurological Institute "acquire international prestige" and he had certainly done that "in the scientific
sphere" because

productivity of the Department began to increase despite the war
. . . and now easily exceeds that of any clinical department in the
University, and of any department of neurology anywhere.[9a]

June 30, 1947: Albert Deutsch, a medical columnist for the liberal
advertisement-free newspaper *PM*, took up Putnam's case, although he stated that Putnam would not discuss the problem with
him.[10] Deutsch attributed the brouhaha to "a powerful combination of big business trustees and a few doctors antagonistic to Putnam." He ascribed the source to an annual deficit of $100,000 and
targeted the question as "whether a voluntary institution tradition-
ally intended to serve mainly those people unable to afford to pay
the full costs of medical care should be put on a 'paying' or self-
supporting basis." He wrote that Putnam had been advised to get
rid of "non-Aryan" physicians (as described in Chapter 6), which
Deutsch considered credible because of "persistent stories" of a
quota system for medical students and faculty. Deutsch also stated
that Rappleye was "said to be favorable to Dr. Putnam but he has
not put up any strenuous fight against the rumor-mongering campaign and it isn't certain he would support him in a decisive battle." In a second article the next day,[11] Deutsch quoted Putnam's
assertion that the hospital was attempting to take over university
functions, including "nomination to staff positions and supervision of research." The third and final article[12] quoted Putnam's
charges that the institute was inadequately maintained, creating a
serious "surgical hazard" in the operating room, and that a nursing shortage was aggravated by directing new nurses to other parts
of the medical center.

July 2, 1947: Putnam wrote to Rappleye on Neurological Institute letterhead, following a conversation the two had had the day before.
Putnam wrote that he had intended to resign from his hospital
positions but wanted to keep his professorships and his salary. He
refused to take either of the positions offered by the Committee
on Administration "because the division of the services constitutes
a clear breach of the agreement under which I left my position at
Harvard and came to Columbia, and which I believe would bring
about the undoing of all I have accomplished in the last seven
years."[12a] Putnam wrote that the October 1945 report of the Executive Committee of the Medical Board and the February 1946
report of the Lamb Committee

constitute a criticism of me. You told me so yourself on June 25, stating that the report of this Committee was considered a complete bar to any honors from the University and Mr. Edward Harkness told me on June 24 that Dr. Lamb had told him there were serious charges against me. As you are aware, I have never heard what these charges were, and have never had an opportunity to reply to them, despite a desperate struggle to bring them into the open.[13]

July 15, 1947: The battle reached the public. The *New York Times* ran a two-paragraph item announcing that Putnam had resigned on July 14.[14] The *Herald Tribune* was more forthcoming, even in its headline: Putnam Quits at Neurological Institute in Row. Director Criticizes Cooper Regime, Cites Poor Maintenance.[15]

How can we make sense of this? The basic problem arises because the hospital kept a paper trail of the legalities but never formally charged Putnam with misdemeanors and never explained fully. Readers are therefore left to speculate.

Putnam himself must have been partly to blame. He had not a single ally. No other service director came to his side; none seems even to have written him a note of condolence. He probably had a formidable foe in Byron Stookey, whose position as head of Neurosurgery he had usurped and who was credited by other neurosurgeons on the staff with having played a role in the resignation of Charles Elsberg, the first chief of Neurosurgery, as a prelude to squashing Putnam.

Putnam's administrative shortcomings are based on hearsay—that he was not rigorous or prompt in setting up the operating room schedule, that he was not diligent in postoperative care of his own patients, and that many of his operations were really research attempts to correct involuntary movements or to deal with problems of the circulation of cerebrospinal fluid.

Looking back now, it was surely an error for the hospital and for Putnam to be director of both Neurology and Neurosurgery. Neurosurgeons do not want to be directed by neurologists and vice versa. No one before or since Putnam has ever tried to be both surgeon and nonsurgeon.

The most direct proof that Putnam was not the right person for the job was the success of his successor, H. Houston Merritt.

Notes

1. Columbia-Presbyterian Medical Center, http://www.smokershistory.com/columbia.htm

2. Dean's Office File. Documentation of conferences re Neurological Institute situation fall of 1944. Archives and Special Collections, A. C. Long Health Science Library, Columbia University.

3. Minutes of the special meeting of the Executive Committee of the Medical Board of the Presbyterian Hospital in the city of New York, held on Friday, October 26, 1945, at 12:00 p.m. Archives and Special Collections, A. C. Long Health Science Library, Columbia University.

3a. Ibid.

4. Ibid.

4a. Ibid.

5. W. C. Rappleye to Frank D. Fackenthal, February 18, 1947. Archives and Special Collections, A. C. Long Health Science Library, Columbia University.

5a. Archives and Special Collections, A. C. Long Health Science Library, Columbia University. Dr. George F. Cahill, chair of the Medical Board, was director of the Urology Service; Dr. Rustin McIntosh was director of Pediatrics; and Mr. John S. Parke was vice president of the Presbyterian Hospital. The director of a service in the hospital is the chair of the corresponding department in the College of Physicians and Surgeons.

5b. W. C. Rappleye to Frank D. Fackenthal, February 18, 1947.

6. Rappleye memorandum, March 5, 1947: Conference with Dr. Tracy J. Putnam. Archives and Special Collections, A. C. Long Health Science Library, Columbia University.

7. T. J. P. to W. A. R., April 11, 1947. Archives and Special Collections, A. C. Long Health Science Library, Columbia University.

8. W. C. R. to T. J. P., April 15, 1947. Archives and Special Collections, A. C. Long Health Science Library, Columbia University.

9. T. J. P. to W. C. R., June 30, 1947. Archives and Special Collections, A. C. Long Health Science Library, Columbia University.

9a. Ibid.

10. Albert Deutsch, Storm over noted neurologist may rock famed medical center. PM, June 30, 1947. Archives and Special Collections, A. C. Long Health Science Library, Columbia University.

11. Albert Deutsch, Medical center heads blocked neurological progress, Putnam says. PM, July 1, 1947. Archives and Special Collections, A. C. Long Health Science Library, Columbia University.

12. Deutsch, Famed neurologist raps medical center as he resigns. PM, July 2, 1947. Archives and Special Collections, A. C. Long Health Science Library, Columbia University.

12a. T. J. P. to W. C. R., June 30, 1947. Archives and Special Collections, A. C. Long Health Science Library, Columbia University.

13. T. J. P. to W. C. R., July 2, 1947. Archives and Special Collections, A. C. Long Health Science Library, Columbia University.

14. Dr. Putnam resigns. Quits his chief executive post at Neurological Institute. New York Times, July 15, 1947. Archives and Special Collections, A. C. Long Health Science Library, Columbia University.

15. New York Herald Tribune, July 15, 1947. Archives and Special Collections, A. C. Long Health Science Library, Columbia University.

9

Phenytoin and the Road from Psychoanalysis to Psychopharmacology

In 1938, when Putnam and Merritt demonstrated the value of phenytoin in treating convulsive disorders, there was only one other effective drug, phenobarbital, which had limited value because it made people sleepy. Bromides had been used but had more annoying adverse effects. For 20 years or so, phenytoin and barbiturates were virtually the only drugs that served as anticonvulsants. Phenytoin not only filled a major gap but also opened the way to the development of other new drugs. By the end of the twentieth century, the standard textbook of pharmacology listed 13 antiepileptic drugs (AEDs).[1] Several of these new drugs have other uses, including the treatment of psychiatric disorders.

Two themes influenced these developments. One was the simultaneous rise of both psychoanalysis and physical or biological treatments for psychiatric conditions. The second theme was the descent of both approaches to treatment with the advent of drug therapy for disorders of mood or behavior. Psychiatrists were central to these changes, but the turmoil involved neurologists, basic neuroscientists, and drug companies.

Neurology and Psychiatry

Although never a matter of contention between Tracy Putnam and Houston Merritt, they must have differed in their views and actions about psychiatry. Putnam's uncle, James Jackson Putnam, was a sponsor of Sigmund Freud's first trip to the United States and is credited with bringing psychoanalysis to the United States, as described in a book by Putnam's great-grandson.[2] In 1894, a year before Freud published *Studies in Hysteria*,[3] William James started a discussion group at the home of the noted Boston psychiatrist Morton Prince. In 1906, Prince founded the *Journal of Abnormal and Social Psychology*, and James Jackson Putnam contributed a paper to the very first issue, describing his experiences with psychoanalysis at the Massachusetts General Hospital. That article was the first public welcome by the medical establishment for Freud's doctrines in America.

Moreover, Tracy Putnam himself was an editor of the journal *Psychosomatic Medicine*. And, for decades, his wife was a practicing psychoanalyst in Boston; she had been trained by Freud himself.

Merritt, in contrast, never expressed his views about psychoanalysis in print; but it was not the kind of data-based medicine he had used in his own studies of neurosyphilis, stroke, and the cerebrospinal fluid. As was his custom, he took a pragmatic view and recognized the clinical practice of the hybrid "neuropsychiatry" in the 1920s and 1930s as an economic necessity for neurologists; they needed the income. He also recognized the use of electroconvulsive therapy by neurologists for its medical indications and as a way to earn money. He welcomed biological psychiatry when it arrived in the 1950s with drug therapy for behavioral and emotional disorders.

It was literally true that some of Merritt's best friends were psychiatrists. Two were especially prominent. Lawrence Kolb was chair of Psychiatry at Columbia during Merritt's long tenure in Neurology there. Francis Braceland had been chief of Psychiatry at the Mayo Clinic before he moved to become director of the Institute for Living, a private psychiatric hospital in Hartford, Connecticut. Both of them visited and stayed at Merritt's summer home in Branford, just outside of New Haven.

A third close friend was Charles Aring, who served with Merritt at the Boston City Hospital, became chair of Neurology at the University of Cincinnati, and was deeply interested in the relations between neurology and psychiatry. Yet another student and colleague of Merritt was Raymond D. Adams, who directed Neurology at the Massachusetts General Hospital from 1951 to 1978 and was himself as influential as Merritt was. Adams had been interested in psychiatry throughout his career and advocated combined training for both specialties, even for trainees headed only for neurology.[4]

Since both neurology and psychiatry were involved in the theory and practice of treating people with emotional or behavioral problems, we might ask how they differ. The answer would probably vary according to different experts. However, the challenge of dealing with altered behavior or mood could be considered psychiatric if no abnormalities are found on the neurological examination. Otherwise, if the "symptoms" (which are described by the patient) are accompanied by "signs" (neurological abnormalities found by the doctor's examination), the physician may conclude that disease of the brain has caused epileptic seizures or abnormalities in alertness, language, movement, or sensation. These are likely to be manifestations of brain dysfunction, so symptoms and signs affecting these skills call for the attention of a neurologist. Psychiatric conditions may cause bizarre behavior, but the physical (neurological) signs of brain damage are not evident, except when there is concomitant dementia.

Similarly, at autopsy, the brain is likely to appear normal in a behavioral disorder. Dementia overlaps both neurology and psychiatry and is manifest by impaired memory, judgment, and language as well as anomalous behavior. Dementia is now usually diagnosed in super-specialized units of departments of neurology. Dementia research is also primarily a field of neurology, including neuropsychology, neuroimaging, or neuropathology. Dictionaries still define the two specialties succinctly. According to one,[5] psychiatry is the "branch of medicine that deals with the diagnosis, treatment, and prevention of mental and emotional disorders." In contrast, neurology is "the medical science that deals with the nervous system and disorders affecting it." As simple as these definitions may seem, they are outdated—perhaps even contentious—because they are based on the concepts of psychoanalysis, which dealt with mind and behavior without a need to consider the brain.[6,7]

Psychoanalysis vs. Physical Treatments: Up and Down Together

At the turn of the twentieth century, mainline psychiatric research was "biological," an approach long dominated by investigators in Germany. They were seeking and failing to find the causes of psychosis by microscopic examination of the brain. Emil Kraepelin (1856–1926), however, was studying living patients, and historian Edward Shorter concluded, "It is Kraepelin, not Freud, who is the central figure in the history of psychiatry."[8]

Kraepelin achieved this high historic status by keeping track of individual patients, recording their symptoms, and following them for years to learn the outcome of the disorder. He recognized the differences in thought and behavior between conditions that later came to be called "schizophrenia" and "manic-depressive disorder." Also, in 1913, he found that 70% of all patients with schizophrenia had a family history of similar symptoms, much more often than in families where no one had the condition.[9] His report was therefore among the first pieces of evidence that genes might be important in these conditions. His description of "senile dementia" was pivotal in what finally became known as "Alzheimer disease."

Kraepelin's belief in a biological origin of psychosis left a lasting mark and presaged the kind of classification that would later emerge in the *Diagnostic and Statistical Manual of Mental Disorders*. Madness is not a single condition.

Kraepelin became the chief of Psychiatry at the University of Heidelberg in 1890. Five years later, he recruited Fritz Nissl and Alois Alzheimer to join him. Nissl was already famed for his development of a stain that identified and could explore abnormalities of nerve cells in microscopic examination of the brain, a technique that is used to this day. In 1906, Alzheimer, another microscopist, found the brain changes that defined the form of dementia that now bears his name. Kraepelin certainly had a knack for recognizing talent in others.

These were lasting contributions, but they did not advance knowledge of either the major psychoses or the "neuroses," the disorders of daily living.

At the same time, Freud was developing his theories of psychoanalysis, which he also used as a treatment. It was soon evident that it could not be used to help people with psychoses. But, for other conditions, it grew in popularity in Vienna and was then imported to North America by physicians who had been banned by the Nazi government from teaching or practice and had to emigrate. Psychiatrists, especially young ones, were attracted to psychoanalysis because, for them, it was a way out of the asylum, which had been their traditional base; additionally, psychoanalysis accommodated an office-based practice, which meant a higher income. The public in the United States was seemingly ready for psychoanalysis in the 1930s and 1940s. Moreover, training in psychoanalysis was abetted by the educational benefits of the G.I. Bill for physicians who were returning veterans of World War II.

The ascendancy of analysis was evident in many ways, none more striking than an episode at Yale University. The chair of Psychiatry there was Eugen Kahn, who had been trained in Germany by Kraepelin himself. In 1948, Kahn was ousted in a "coup by young analytically-oriented psychiatrists."[10] By 1950, only 10% of all psychiatrists were analysts but more than half of the chairs of psychiatry departments in the United States were analysts.[11]

The rise of psychoanalysis was accompanied by a parallel rise of a biological approach that was based on theories of brain dysfunction, which led to physical treatments aimed at healing the brain. One of the first of these "biological" treatments was recognized in 1927, when the Nobel Prize was given to the Viennese neuropsychiatrist Julius Wagner-Jauregg (1857–1940). He had developed an effective treatment for "general paresis," the dementia of neurosyphilis, by giving the patient a fever.[12] At first, this was done by giving an intravenous injection of the organisms that cause malaria. However, malaria treatment could not be standardized; it was difficult to decide how many malaria organisms to give in how many minutes for patients who differed in responsiveness. Malaria treatment was therefore replaced by the more predictable and more reliable electrical "hot box" to induce the fever. The patient was encased in a metal box, with only the head protruding. Then, large electric bulbs were turned on to generate sufficient heat to raise the body temperature, giving the subject a fever. It seemed to be effective and was used for decades through the time of World War II. After the war, penicillin came along and almost eradicated all syphilitic infections; cerebral syphilis disappeared, and, even though the advent of HIV infection and AIDS fostered the reappearance of other sexually transmitted diseases, cerebral neurosyphilis never again became a prevalent scourge in advanced countries.

During this period, four physical treatments were also popular: frontal lobotomy, insulin coma therapy (which often led to convulsions), and two types of convulsive therapy, one induced by a drug, pentylenetetrazol (Metrazol), and the other by electrical shock. These treatments were administered by "biological psychiatrists," and some of them were critical of psychoanalysis. Others were just focused on the biological approach. One such was Lothar Kalinowsky (1899–1992), who was only half Jewish, but that sufficed to make him a quadruple emigrator. He received his MD in Berlin in 1922, left Germany for Italy in 1933, and had to return to medical school for a second degree in 1933 so that he could be certified in his new country. In Rome, he worked with Ugo Cerletti, who is credited with the introduction of electroconvulsive therapy (ECT) in 1938, following years of less controllable coma induced by insulin or convulsions that followed injections of Metrazol.[13]

With the arrival of the Germans in Rome, the Mussolini fascist government restricted the lives of Jews, and Kalinowsky was off again. In 1939, he took ECT to Paris, where it rapidly replaced Metrazol convulsive therapy. He kept moving and, in 1940, landed at the New York Psychiatric Institute, a division of both the state government and Columbia University. There, within six months, he had established an ECT unit. He was later appointed to the staff at the neighboring Neurological Institute and other hospitals in New York, where he had a flourishing practice as one of the leading figures in ECT. Kalinowsky was ultimately deemed "the grandparent of American ECT."[14] A consensus conference of experts was convened by the National Institutes of Health (NIH) and endorsed the treatment in 1985;[15] it is still used now, especially for the treatment of severe and persistent depression.[16]

Concepts about brain and mind were already changing in the 1930s when psychoanalysis was on the rise, and after World War II analysts had a dominant role in academic departments of psychiatry. "Psychopharmacology" is often said to have started in 1952 with the serendipitous discovery that chlorpromazine (Thorazine) had a calming effect on patients being prepared for anesthesia and surgery.[17,18] Then, isoniazid was accidentally found to have antidepressant properties when it was given to enhance antibiotic therapy for tuberculosis. That experience led to the synthesis of similar drugs, which culminated in the 1958 introduction of imipramine (Tofranil). A third chance discovery was made by John Cade, who was studying the effects of uric acid, a normal chemical found in human tissues, blood, and urine. He had difficulty preparing solutions for intravenous use in animal experiments and found that the addition of lithium salts corrected the problem. More, it had a sedating effect on guinea pigs. He published his results in 1949 in an obscure Australian journal, but it was only in 1970 that the Food and Drug Administration (FDA) gave approval for the treatment of mania or bipolar conditions.[19]

Both psychoanalysis and physical treatments faded in the 1950s and 1960s with the introduction of drugs to treat psychiatric disorders—antidepressants and effective medication for schizophrenic symptoms, mania, anxiety, and panic attacks. Neuroscience had also progressed, and drugs were designed to act on specific chemical transmitters of the brain. Some worked on serotonin; others on dopamine. Abnormalities in different brain chemicals caused different clinical symptoms, and different drugs were needed. These new drugs and chemical advances changed concepts of the fundamental nature of mental diseases. If the clinical problems arise from faulty transmission of nerve impulses from one brain cell to another, that would be a chemical problem. If so, how could talk therapy work? Moreover, it has been difficult to prove the efficacy of psychoanalysis as a therapy because it is so difficult to carry out a double-blind controlled trial, in which treatment is compared with a placebo; that is possible with therapeutic drugs.[20]

Drug therapy also emptied the asylums. Peak occupancy in state mental hospitals was 559,000 in 1955. In 1970 the figure was 338,000 and in 1988, 107,000, a decrease of 80% in 30 years.[21]

No wonder that psychoanalysts disappeared from academic leadership in psychiatry, but their influence is still considerable.[22-24] Eclectic psychotherapy is still in wide use, and the number of American analytic training programs increased from 20 in the 1960s to 29 after 2005.[25]

Merritt's Views

Merritt's basic consideration was consistent through the years. Neurology should be a discipline of its own, not tied to psychiatry, neurosurgery, or internal medicine.

In the mid-1960s, when psychoanalysis was declining, some younger neurologists advocated a separation of the certifying board examinations. Until then, young neurologists and psychiatrists had to pass an oral examination that covered both fields. The renegades considered this a silly waste of time and effort. At a national meeting, Merritt called attention to the paucity of neurologists and the plethora of psychiatrists in the United States. Neurologists, he pointed out, are too few to have any political impact by themselves but, in concert with psychiatrists, they might have some influence. The ensuing unofficial ballot at that meeting was overwhelmingly in favor of Merritt's views.

Nevertheless, in 1975, Merritt wrote a history of neurology for the American Neurological Association and included a discussion of the two specialties. He wrote that 50 years earlier "all practitioners of the specialty were neuropsychiatrists." They took

> pleasure from the study and treatment of patients with organic dis-ease of the nervous system but (they made their) living by the treat-ment of psychiatric patients. There were only a few residencies in neurology... Even as late as 1936, there were only 16 hospitals listed in the United States as having approved training for residency in neurology. In addition most of the physicians who took a residency in neurology went on to practice neuropsychiatry.[26]

He claimed that training was so skimpy and so brief, usually one year, that most of them "did not know much more about diseases of the nervous system than did the internists who received several years of training in internal medicine." He concluded that 50 years earlier, in 1925, "clinical neurology was almost moribund."

He attributed the problem partly to the rapid growth of neuro-surgery and psychiatry in the 1930s and 1940s. "There were only two or three full-time appointments in neuropathology that were not in psychiatric institutions," he lamented. Neurology did advance, and he thought that, with the development of modern neurosciences, neurology has "made enormous strides and has become a specialty of primary importance in the medical sciences."

However, the true impetus for the development of clinical neu-rology came with the establishment of the National Institute of Neu-rological Diseases and Blindness in 1948.[27] Within a generation, the number of residency positions increased from 32 to more than a 1000 in 106 hospitals. Additionally, basic neuroscience and patient-oriented research grew rapidly when NIH research training and pro-gram grants became available.

Phenytoin played a role in this transformation. One year after their report of the antiepileptic efficacy of phenytoin,[28] Putnam and Merritt described episodic behavioral problems, which they called "dulness"[29] [sic] and which could arise in different ways. First, drowsiness or cloud-ing of thought could be an adverse effect of an anticonvulsant drug.

Second, anticipation and fear of the next attack might interfere with concentration and "psychologic conflicts may form a part of the chain of events leading to an attack." Third, "psychomotor attacks" were already recognized as manifestations of seizures arising in the temporal lobes; these attacks might be characterized as peculiar and repetitive motor behavior for which the patient later had no recollection. Sometimes, these behavioral anomalies were the only evidence of an attack.

The episodes of concern in that paper were of a different nature, "a disturbance of intellectual function in which the entire attack remains subliminal, unnoticed by either the patient or those about him and of course unaccompanied by unconsciousness or motor or sensory disturbance." They noted that these "absences" might occur several times a day without impeding work or study or the effects could be "crippling to any complicated form of mental activity."

They found that these attacks were associated with transient runs of abnormal cerebral activity in the electroencephalogram. That is, the attacks were seizures without convulsions.

Today, anticonvulsant drugs are used widely in the treatment of psychiatric disorders, especially bipolar conditions and schizophrenia. Putnam and Merritt did not sustain interest in these applications, but in their 1938 paper on the efficacy of phenytoin as a treatment for epilepsy, they wrote:

> In addition to a relief or a great reduction in the frequency of attacks it was frequently noted by the parents of children that they were much better behaved, and more amenable to discipline and did better work in school.[29a]

In 1943, Kalinowsky and Putnam reported improvement of manic symptoms with phenytoin treatment, but the study was not controlled for placebo effects.[30] Two generations later, a controlled study confirmed the benefit.[31] Others noted effects on mood in children treated for seizure disorders with phenytoin.[32]

However, the major impetus for the psychiatric application of AEDs came in the Decade of the Brain, 1990–2000. That declaration was a public relations term intended to attract the interest of Congress to increase the NIH research budget by sending an explicit message: "Wondrous medical applications of the advances in fundamental neuroscience are coming and this prospect warrants more research funds—now." During those years, eight new AEDs were introduced.[33]

When the newer AEDs began to appear, they were used "off-label." That term means the drugs were first deemed safe and effective for epilepsy by the FDA and, being safe for human use, physicians could prescribe them for purposes other than those approved by the FDA. In the 1960s, carbamazepine (Tegretol) was found to be effective in treating intermittent facial pain (trigeminal neuralgia), which shares with epilepsy the bursting and ever-surprising quality of being "paroxysmal." The nonepileptic conditions treated with these drugs include abnormal heart rhythms, shingles and the post-herpetic neuralgia that follows it, and diverse psychiatric symptoms.[34–39]

One of the new drugs was carbamazepine, which had been produced in the laboratory by modifying the chemical structure of imipramine, one of the early antidepressant drugs. Carbamazepine was therefore an AED that was used by psychiatrists from the start. By 2000, these drugs were in widespread use by psychiatrists. Neurologists were also using them for nonepileptic disorders, including pain syndromes, migraine, and tremor.[40,41]

The Strange Public Case of Financier Jack Dreyfus: Advocate for Phenytoin

The most ardent proponent of the widespread use of phenytoin in psychiatry was Jack Dreyfus, inventor of the mutual fund for personal investors. He became the "Lion of Wall Street" and one of the richest men in the United States but still seriously depressed in 1957, at age 42. His personal physician, one of the leading internists at Presbyterian Hospital, recommended Maximilian Silbermann as neuropsychiatrist. One of the six German-Jewish refugee neuropsychiatrists hired by Putnam, Silbermann had a kindly demeanor; and although he was not a psychoanalyst, he and Dreyfus engaged in prolonged psychotherapy, meeting five or six times a week for several years.

After about 6 months, Dreyfus was admitted to the Harkness Pavilion, the private wing of Presbyterian Hospital. He was not suicidal and left after a few days. Silbermann rejected psychoanalysis when it was suggested by family members. He proffered no specifics about the possible cause but suggested something was wrong with body "chemistry." Somehow, Dreyfus linked this to body "electricity," and he mentioned this to Silbermann, who had told him that epilepsy was considered an electrical disorder. Dreyfus recalled a girl he knew who had epileptic attacks that were controlled by medication. He asked what that could have been, and Silbermann responded, "Dilantin."

"Why don't I try that?" asked Dreyfus.

Silbermann responded, "You can try it if you like. I don't think it will do you any good, but it won't do you any harm."

Dreyfus recovered rapidly, and the psychotherapy sessions were terminated. He began to talk to others who had depressive symptoms and told them to seek phenytoin therapy from their doctors; in the process, he accumulated anecdotes that supported his trust in the drug. Seemingly, he encountered no failures or adverse reactions.

Dreyfus was so upbeat that he gave Silbermann a Rolls Royce. He also welcomed a suggestion from Silbermann. He should invite Merritt, the discoverer of Dilantin, to dinner. Merritt came and brought with him Lawrence C. Kolb, chief of Psychiatry at Columbia-Presbyterian Medical Center. Several other physicians and friends were there, too. Over a fine private dinner, Dreyfus recounted his series of five personal cases and quoted from a textbook of pharmacology that the drug had "salutary effects on personality, memory, mood, cooperativeness, emotional stability, amenability to discipline, etc. . . . sometimes independently of seizure control." Merritt assured them that the safety record was good. He told Dreyfus that what was needed was a formal, blinded, controlled study of the drug.

A few days later, Kolb called to report that a young psychiatrist, Sidney Malitz, would be responsible for setting up the clinical trial. There followed a series of communication failures, and Dreyfus blamed the problems on Malitz's delays in responding. In turn, Malitz blamed the delay on problems in obtaining placebo pills from the manufacturer. Dreyfus became impatient, and from his own writings, it was clear that he did not fully understand the importance of placebo controls; thus, attempts to set up the trial were terminated in 1965.

In a well-designed therapeutic study, patients are divided into two groups, in numbers sufficient to detect a statistically significant difference in response to treatment. Patients in the treatment group are given doses of the authentic drug. Those in the control group are given "placebo," an inert substance in pills configured to look identical to the test compound. Patients are selected to meet strict diagnostic criteria for the condition under study and are matched for similarity in age, sex, and duration of symptoms or other characteristics. They are assigned randomly to one group or the other. The trial is "blind" because neither the patients nor the treating physicians know which treatment any individual is taking. Often, a third doctor evaluates the results for safety, also without knowing what the treatment is for a specific patient. The code is broken if any subject has an adverse reaction. The trial is set with those restrictions to assure that improvement cannot arise from the physician's power of suggestion, the patient's wishing well, or any other influence except for the drug being tested.

That is not the way Jack Dreyfus ran his trials. Instead, he spoke to a friend or acquaintance, determined himself whether the person was depressed, and suggested treatment with phenytoin. A few days later, he would ask whether the subject felt better, and almost always the answer was affirmative.

The following discussion is from the written report of a taped discussion of Jack Dreyfus with friends and advisors;[42] "PHT" stands for phenyl hydantoinate, a chemical name for Dilantin.

Jack: Yura, we are talking about research, right? Please listen before you say no. None of these people who took PHT knew each other. As far as they were concerned the study was blind. I asked them to write me letters that included details of their experiences. The same results from PHT are reported over and over again. This reinforces the evidence.

Dr. Peter S: It is not accepted as proof and there's a devastating word that is applied to it, called anecdotal evidence. It doesn't go.

Yura: It's indirect proof.

Jack: Sorry fellows. Nobody in the room is thinking. These individuals wouldn't know which way to lie if they wanted to. They didn't know each other.

He just did not understand the nature of a controlled trial. Years after being publicly criticized by Dreyfus, Sidney Malitz had become well known as a research psychiatrist and was twice acting chair of the distinguished department at Columbia University. In an interview, he said that Dreyfus "wanted us to come up with an answer he wanted whether he realized it or not." Malitz did a preliminary uncontrolled trial on "a sufficient number of patients that convinced us that Dilantin was not promising."[43] After others attempted to do a few desultory therapeutic trials, Dreyfus himself carried out a study of jailed men who had committed crimes and were not being treated for specific mental illness, again without appropriate design. Jailed men cannot really give informed consent for an experiment if the test leads to more lenient conditions.

Dreyfus set up a research foundation and claimed that he spent more than $80 million in the next two decades, a sum that would be equal to much more today in current dollars.[44] Much of the money went for publication and distribution of three books.[17,45,46] The first book was published by Simon and Schuster, but the $2 million marketing campaign was paid by Dreyfus himself.[47] He hired Samuel Bogoch, a neuroscientist at Harvard, to serve as medical advisor for the jumbo bibliography of papers mentioning phenytoin and emotional disorders, which was mailed to 150,000 physicians throughout the United States.[47a] Despite his relentless efforts,[48,49] he never convinced the FDA to approve phenytoin for the treatment of psychiatric disorders.

Despite his failures, phenytoin and other AEDs have been used effectively to treat irregular heart rhythms. Carbamazepine (Tegretol), another AED, has been used to treat paroxysmal facial pain in trigeminal neuralgia, and AEDs are in wide use to treat psychiatric disorders. Was Jack Dreyfus deluded, visionary, or both?

Notes

1. McNamara JO. Pharmacology of the epilepsies. In Brunton LL, Lazo JS, Parker KL, eds. Goodman & Gilman's The pharmacological basis of therapeutics, 11th ed. New York: McGraw-Hill, 2006, p 509, table 19–2.
2. Prochinik G. Putnam camp. New York: Other Press, 2006.
3. Shorter E. A history of psychiatry. New York: John Wiley & Sons, 1997, p 167.
4. Price BH, Adams RD, Coyle JT. Neurology and psychiatry: closing the great divide. Neurology 2000;54(1):8–14.
5. American Heritage Dictionary of the English Language, 4th edition. Boston: Houghton Mifflin, 2000.
6. Eisenberg L. Mindlessness and brainlessness in psychiatry. Br J Psychiatry 1986;148:497–508.
7. Eisenberg L. Is psychiatry more mindful or brainier than it was a decade ago? Br J Psychiatry 2000;176:1–5.
8. Shorter, A history of psychiatry, 100.
9. Ibid., 240.
10. Ibid., 171.
11. Eisenberg L. Past, present, and future of psychiatry: personal reflections. Can J Psychiatry 1997;42:705–713.

12. Healy D. The creation of psychopharmacology. Boston: Harvard University Press, 2002, p 51; http://en.wikipedia.org/wiki/Julius_Wagner_von_Jauregg
13. Shorter, A history of psychiatry, 218.
14. Ibid., 221.
15. Runck B. NIMH report. Consensus panel backs cautious use of ECT for severe disorders. Hosp Community Psychiatry 1985;36(9):943–946.
16. Greenberg RM, Kellner CH. Electroconvulsive therapy. A selective review. Am J Geriatr Psychiatry 2005;13(4):268–281.
17. Bhatara VS, Lopez-Munoz F, Gupta S. Celebrating the 50th anniversary of the introduction of chlorpromazine in North America and the advent of the psychopharmacology revolution. Ann Clin Psychiatry 2005;17(3):108–111.
18. Domino EF. History of modern psychopharmacology: a personal view with emphasis on antidepressants. Psychosom Med 1999;61:591–598.
19. Shorter, A history of psychiatry, 258.
20. Paris J. The fall of an icon. Psychoanalysis and academic psychiatry. Toronto: University of Toronto Press, 2005.
21. Shorter, A history of psychiatry, 280.
22. Kernberg OF. Psychoanalytic contributions to psychiatry. Arch Gen Psychiatry 2002;59:497–499.
23. Wallerstein RS. Psychoanalytic treatments within psychiatry. Arch Gen Psychiatry 2002;59:499–500.
24. Kandel ER. Biology and the future of psychoanalysis: a new intellectual framework for psychiatry revisited. Am J Psychiatry 1999;158:505–524.
25. Shorter, A history of psychiatry, 171; and 2006 figure from the American Psychoanalytic Association website, http://www.apsa.org
26. Merritt HH. The development of neurology in the past fifty years. In Denny-Brown D, ed. Centennial Annual Volume of the American Neurological Association, 1875–1975. New York: Springer Publishing Company, 1975, pp 3–10.
27. Rowland LP. NINDS at 50. NIH Publication 01–4161. 2001; also New York: Demos Medical Publishing, 2003.
28. Merritt HH, Putnam TJ. Sodium diphenyl hydantoinate in the treatment of convulsive disorders. JAMA 1938;111:1068–1073.
29. Putnam TJ, Merritt HH. Dulness (sic) as an epileptic equivalent. Arch Neurol Psychiatry 1941;45:797–813.
29a. Ibid.
30. Kalinowsky LB, Putnam TJ. Attempts at treatment of schizophrenia and other non-epileptic psychoses with Dilantin. Arch Neurol Psychiatry 1943; 49:414–420.
31. Mishory A, Yaroslavsky Y, Bersudsky Y, Belmaker RH. Phenytoin as an antimanic anticonvulsant. Am J Psychiatry 2000;157:463–465.
32. Lindsley DB, Henry CE. The effect of drugs on behavior and the electroencephalograms of children with behavior disorders. Psychosom Med 1942;4:140–149.
33. Bazil CW. New antiepileptic drugs. Neurologist 2002;8:71–81.
34. Amols W. Differential diagnosis of trigeminal neuralgia and treatment. Headache 1969;9(1):50–53.

35. Nadkarni S, Devinsky O. Psychotropic effects of antiepileptic drugs. Epilepsy Currents 2005;5(5):176–181.
36. Ovsiew F. Antiepileptic drugs in psychiatry. J Neurol Neurosurg Psychiatry 2004;75(12):1655–1658.
37. Muzina DJ, El-Sayegh S, Calabrese JR. Antiepileptic drugs in psychiatry—focus on randomized controlled trials. Epilepsy Rev 2002;50:195–202.
38. Golden AS, Haut SR, Moshé SL. Nonepileptic uses of antiepileptic drugs in children and adolescents. Pediatr Neurol 2006;34(6):423–432.
39. Zaremba PD, Bialek M, Baszczyk B, Cioczek P, Czuczwar SJ. Non-epilepsy uses of antiepileptic drugs. Pharmacol Rep 2006;58:1–12.
40. Rogawski MA, Löscher W. The neurobiology of antiepileptic drugs for the treatment of nonepileptic conditions. Nat Med 2004;10:685–692.
41. Ovsiew, Antiepileptic drugs in psychiatry.
42. Bhatara, Lopez-Munoz, Gupta, Celebrating the 50th anniversary, 154.
43. Sun M. Book touts Dilantin for depression but researchers and the drug's manufacturer deny Dilantin's value as an antidepressant. Science 1982;215(Feb 19):951–952.
44. Bhatara, Lopez-Munoz, Gupta, Celebrating the 50th anniversary, 137.
45. Dreyfus J. A remarkable medicine has been overlooked. New York: Simon & Schuster, 1981. Also published by Dreyfus Medical Foundation, New York, 1981, 1982.
46. Smith BH, Bogoch S, Dreyfus J. The broad range of clinical use of phenytoin: bioelectric modulator. 1988, New York: Dreyfus Medical Foundation.
47. Sun, Book touts Dilantin.
47a. Ibid.
48. Dreyfus J. The story of a remarkable medicine. New York: Lantern Books, 2003, pp 117–128.
49. Bhatara, Lopez-Munoz, Gupta, Celebrating the 50th anniversary, 187.

He is truly great what is little in himself and that maketh no account of any height of honors.

Thomas à Kempis[1]

At the fiftieth anniversary celebration of the American Neurological Association in 1975, Merritt was celebrated. "After a dramatic pause as the last of the awards was given, the Chairman, Dr. Joseph Foley, told of his pleasure and honor of giving the next award, which had been saved till last—it was for greatness and was given to 'H. Houston Merritt, Neurologist to the World.'"[1]

10

H. Houston Merritt, 1902–1979

Houston Merritt was not a Boston Brahmin. He was not born into any upper class. He was proud to be a Tar Heel, a native of North Carolina. He was born in Wilmington, the largest city of that state and the largest port but no longer a great metropolitan area; the population in the year 2000, was 75,838.

Wilmington was famous for another reason, the "race riot" of 1898, in which black leaders were massacred.[2] The population then was 20,055. Blacks not only outnumbered whites but also held responsible positions in business and government. In the melee, whites massacred blacks in uncertain numbers. Estimates from 10 to 300 deaths were recorded. Within days, a massive exodus of blacks changed the racial balance in Wilmington, Reconstruction ended, and Jim Crow was installed throughout the southern states.

However dramatic that event was, Houston Merritt was born four years later, and there is no record that it affected him one way or another. Nothing in the meager documentation of the family history mentions the riot.

In 1968, Edward Merritt and Leora Hiatt McEachern, husband and wife members of the Merritt family, compiled a listing of their family starting in the mid-nineteenth century (some 300 years after the arrival of the Putnams in Massachusetts). The first Hiram Houston Merritt was born in 1870 and died in 1945.

Hiram, Sr., was a midlevel manager for a local railroad company. He married Dessie Ella Cline; their religion was Methodist.[3] They had a daughter, Lucille, and two sons, Harry and Hiram Houston, Jr. The younger man gave his name as H. Houston Merritt.

He attended public schools, graduating from Wilmington High School in 1918, at age 16. As an undergraduate at the University of North Carolina, he came to public attention, as a newspaper headline told the story: Wilmington Boy Wins High Honor at N.C. University; H. H. Merritt, Jr. Among Eight Making Highest Possible Grade.[4] The highest possible grade was "above 95%," an achievement attained by seven sophomores and Houston, the only freshman.

Brother Harry became the "cashier" for the Pacific Redi-Cut Houses Corporation in California. When his parents traveled to see their son, the novelty of the trip sufficed to warrant a newspaper item.[5] During World War II, Harry was first an Army captain and then a major, serving in Germany.[6]

Although he spent only one year at the University of North Carolina in Chapel Hill, Houston felt a serious attachment; much later, in 1979, he endowed a professorship of neurology there. After that first college year, Houston resumed his accelerated career, transferred to Vanderbilt University, and graduated in two years. After only three years of undergraduate study, he went to medical school at Johns Hopkins, graduating in 1926 at age 24 and moving to Yale for an internship in medicine.

Merritt's experience at Yale affected the author. When, many years after Merritt had been at Yale, 1950 it was, I was a medical student as well as a medical intern there. When it was time for me to apply for neurology training, I needed a letter of recommendation. A family friend was Harry M. Zimmerman, a renowned neuropathologist, so I asked him whether he would write a letter on my behalf. Zimmerman replied that he would be pleased to write because, he said, "Houston and I are good friends. When I was an Assistant Resident in Medicine, he was my intern."

When the day came for my interview with Merritt, he was reading my application form and the enclosed letter. He commented, "I am pleased to see that you have a letter from Harry Zimmerman. You know, he and I were very good friends. When I was an Assistant Resident in Medicine, he was my intern."

Back to Merritt's story. In 1928, he moved to the Boston City Hospital (BCH) for training in neurology. Stanley Cobb was the chief and led what was probably the most distinguished program in the country that attracted many talented people. Many of them were actually preparing for careers in academic psychiatry, not neurology; but there was little distinction between the two disciplines. Cobb remained a neurologist at BCH until 1934, when he left to create the first psychiatry department at the Massachusetts General Hospital. As described in Chapter 3, those were glory years for BCH, the birthplace of modern clinical neurology and the cradle of clinical investigation in neurology, internal medicine, and surgical specialties.

Medical specialty training in 1928 was not the highly organized five-year ritual and obstacle track it is now. Specialty boards were ultimately organized to set training standards and examination requirements, but they were not created until 1934. There were few training programs and few candidates when Merritt came along. Not even the New York Neurological Institute was known for training. Today, a computerized "matching" system makes the complicated system of application and selection a reasonable and fair process.

Merritt met Mabel Carmichael when they were both working at New Haven Hospital; they married in 1930. They had no children, but they were a durable couple for 49 years, separated only by his death. She became part of his life, mirroring his direct and plain-speech personality, often inviting residents to their home for dinner and poker. She was involved in the dinner but not the card game. She was with him for every public function. They had a summer home in Branford, a suburb of New Haven on Long Island Sound. There, they welcomed visitors, current or former residents, or houseguests for old-timers like Charles Aring or Frank Forster.

After his residency, Merritt took the usual academic tour to Europe; the voyage was their honeymoon. As a fellow of the National Research Council, he opted to work with Walther Spielmeyer, a leading neuropathologist in Munich, and became interested in neurosyphilis. Hitler was coming, but Merritt never spoke of it to his associates. As a matter of fact, he rarely commented on politics, national or international; but he was less reserved about institutional affairs.

While Merritt was away, the Neurological Unit at Boston City Hospital (BCH) was founded, under the direction of Stanley Cobb, who immediately had serious administrative problems. According to Cobb's biographer,

> Cobb had increasing friction in relations with John J. Dowling, the administrator of BCH. In April 1929, when ground was finally broken for the new neurological unit, a situation arose which Cobb viewed as a crisis. An important aspect of Cobb's strategy in setting up the neurological unit was to incorporate neurosurgery, and

Dowling had assured Cobb verbally that two operating rooms and other needed facilities for neurosurgery would be available on the 8th floor. However, Dowling was unwilling or neglected to give this assurance in writing.—Cobb threatened to pull out as he wrote in a letter to Dean Edsall. According to Byrne, "Cobb continued to have trouble with the hospital administration…as an ongoing frustration."[7]

On returning to Boston in 1931, Merritt was appointed to the lowest academic level at Harvard. Gradually, he assumed higher titles and was a full professor by the time he left in 1944. Throughout this period he had an appointment at BCH.

Merritt's clinical and scientific papers started to appear in 1930. The first was written with Merrill Moore, who later achieved fame as a poet. Their subject was a form of nerve injury.

Merritt rapidly became a consummate clinician. He relished every opportunity to solve a diagnostic problem. His colleagues at BCH were all involved in epilepsy research, following Cobb's lead. And Merritt would soon join them, but in the meantime, he was the one to be called to see patients at any of the Harvard hospitals, including Mass General and the Brigham. That included Saturdays and Sundays.

In 1934, Cobb left BCH to set up the first psychiatry service at Mass General. Tracy Putnam took his place, and by 1935 or 1936 he and Merritt had joined forces to "make a systematic survey" of antiepileptic drugs, the first step on the way to the discovery of Dilantin.[8] From that time onward, until the very end of his career (and his life), epilepsy was central to Merritt's research and practice.

In the meantime, Putnam himself left BCH in 1939, apparently lured primarily by the possibility of a larger research budget at the New York Neurological Institute but probably also pushed by problems with the leading administrator at BCH and possibly by problems with Donald Munro.

By the time Putnam left, Merritt had established himself as the popular favorite to take over, as documented by several published comments. Merritt's appeal was not limited to professionals but included the supporting staff at BCH, many of them Irish immigrants or descendants.

Merritt was appointed interim director but was passed over when the final decision was made. His main opponent, it turned out, was none other than the man who had attracted him to BCH in the first place, Stanley Cobb. Another Brahmin, Cobb considered Merritt too crude. "However, probably because of what Cobb perceived as his lack of social grace, he was not even considered for the permanent position."[8a]

According to Joseph Foley,[9] "Cobb was also a Brahmin and disdained crudity in speech or manner—so he did not like Houston and made that clear." Cobb also told Foley that Catholics could not be good physicians.

Benjamin White, Cobb's biographer[9], on the same page, also wrote that "Merritt was a rough diamond socially and even during the interview in late life had a somewhat gruff manner."[9a] That interview was carried out in 1977 by this author; Merritt was approaching the peak of disability from what would be his fatal illness and was manifestly uncomfortable throughout the recorded conversation. (The tape was lost, later found by Robert Kurtzke in the trash when the Neurological Institute library was being pruned, and finally given to Michio Hirano for safe keeping before being returned to the author. It has been deposited in the Special Collections division of the College of Physicians and Surgeons library.)

In the words of Francis Forster, when Putnam resigned, "It was almost universally expected that Merritt would succeed. Instead, the Harvard Trustees appointed Dr. Derek Denny-Brown."[10] Only weeks later, in 1939, World War II commenced with the German attacks on its neighbors.

Denny-Brown (1901–1981) was only one year older than Merritt, but they were as different as any pair of neurologists could be. Born in New Zealand, Denny-Brown graduated from Dunedin medical school in 1924 and moved the next year to Oxford, where he worked with Sir Charles Sherrington, a Nobel laureate. There, he became an accomplished neurophysiologist but wanted to continue clinical research as well. In 1931, he therefore went to the peerless National Hospital in London for clinical training. His bibliography includes a striking mix of physiology, neuropathology, and clinical neurology. Few other neurologists have ever made so many contributions in so many different fields of medical endeavor.

In 1939, Cobb was seeking a replacement for Putnam at BCH. He and the president of Harvard, James Bryant Conant, invited Denny-Brown to Boston for an interview. The decision was made then and there; they offered him a position as chief of Neurology at BCH and a

Harvard appointment as professor. But, just as that happened, the war intervened. Harvard announced that the position at BCH would be held for Denny-Brown, who was called to serve in the British army at a military center for head injury in Oxford. His research now focused on the physiology of concussion and how it might be prevented or the brain injury ameliorated. "Concussion" is defined as a brief loss of consciousness after a head injury without other signs of brain injury on neurological examination. However, even momentary loss of consciousness could jeopardize a soldier's performance or even his life.

Merritt was appointed interim director of the unit and awaited the arrival of Denny-Brown in 1941; Merritt served as an assistant to Denny-Brown until 1944. In 1941, the United States was not yet a participant in the war; but Conant spoke to Winston Churchill, and arrangements were made for Denny-Brown to come to the United States with high priority. He was made a member of the federal Committee for Aviation Medicine in Washington, D.C., while simultaneously serving as a Harvard professor, leading the Neurological Unit at BCH, studying nerve injuries in the laboratory, and adjusting to a new culture in Boston. Fortunately, the director of BCH was not the one who had hounded Putnam but a person who was friendlier to the unit.

Four years after his arrival in Boston, in 1945, Denny-Brown was called to service in the British army in India and Burma. While there he wrote a paper about neurological disorders, primarily caused by malnutrition, in prisoners of war. He returned to Boston in 1946 as the James Jackson Professor of Neurology. By that time, Merritt had departed to accept a position as chief of Neurology at the Montefiore Hospital, a teaching service of Columbia University, with the title of professor.

Merritt worked under Denny-Brown from 1941 to 1945. We can only imagine how they got along. Merritt was affable, intuitive, unflappable, and of modest origins. Denny-Brown had the hauteur of British nobility (which he was not) and, to boot, a hyphenated name. His admirers wrote that at professional meetings "he was known as a formidable discussant of the scientific presentations of others, frequently highly critical and incisive, and at times devastatingly frank." They remarked that he was "given to outbursts of anger if house officers transgressed by presenting faulty information."[11]

Even their literary styles differed, although both were prolific writers. Merritt was succinct and direct; Denny-Brown was flowery and erudite. According to one apocryphal story, a famous British neurologist wrote to him, "Dear Denny, I see you have a paper in *Brain*. When is the English version coming out?" (*Brain* here is the name of a leading neurology journal.)

One of Denny-Brown's special interests was in the "basal ganglia," structures deep within the brain that are involved in the control of movements. When these structures are struck by disease, involuntary movements result; manifestations of these diseases include the tremor of parkinsonism and the dancing-like movements of Huntington chorea.

According to the inimitable Joseph Foley, an illustrious BCH alumnus who later directed Neurology at the Metropolitan Hospital in Cleveland, Denny-Brown's problems were extensive.

[He] still had to deal with elevator operators who wouldn't stop at his floor, with perverse Bostonians who insisted on calling him "Dr. Brown," and with the patronizing airs of that small group of faculty who for over a hundred years never really accepted the Boston City Hospital as part of Harvard Medical School.[11a]

How did Merritt and Denny-Brown get along in those years? Merritt never made a public statement that might raise doubts about their collegiality. But I heard two stories. Frank Forster said that at a teaching session one day Denny-Brown was examining a patient's position sense. Merritt became impatient and rose to show Denny-Brown how to move the patient's toe without giving physical clues to the direction of movement. According to another story[11b], Denny-Brown had lost the left ring finger, which was surgically amputated because he had developed a "Dupuytren contracture," a fixed posture of flexion of the finger that is attributed to changes in connective tissue for reasons unknown. That occurred when Denny-Brown was still at the National Hospital, Queen Square, in London. In his typical inquisitive fashion and the British tradition of self-experimentation, he performed an experiment on himself. He had the digital nerves crushed about a week before the planned surgical amputation. He then studied the expected development of increased skin sensitivity to light touch in the distribution of the crushed nerve. After the amputation, he and other colleagues studied the excised nerves under the microscope, where he found that newly sprouting nerve fibers arose from the crushed nerve and that these new fibers could account for the abnormally painful sensations. It seems likely that Denny-Brown suffered from either severely intense paresthesias or pain. The perception of pain from stimuli that do not ordinarily result in pain even has a medical name, "allodynia." Here, it was presumably related to the hypersensitivity of the newly formed sensory nerves. Denny-Brown found some surcease by rubbing the stump of the missing or "phantom" finger. As a result, he did indeed rub the proximal portion of the amputation site with his good right hand.

John Meyer, one of the first to use cerebral blood flow studies for clinical purposes, was a trainee of Denny-Brown's in 1950–1953, starting six years after Merritt had left BCH. He contrasted the styles of the two, who

> were entirely different as neurologists. HHM took a careful history, did a brief neurological examination and his clinical diagnosis was usually correct, whereas DDB relied more on neurological signs so his clinical diagnosis was more often wrong, but brilliantly so. DDB applied his knowledge of neurophysiology and neuropathology to clinical neurology and put his deductive reasoning to original solutions of clinical neurological problems. He was more of an experimental clinical neurologist.[12]

Meyer contrasted the two even more succinctly. Merritt was "an amusing southern gentleman with a sense of humor, who listened to others and made excellent group and committee decisions." Denny-Brown was "not a good group leader, was impatient and not well-liked, although brilliant."[12a]

Professor Sid Gilman, however, tells a different story, based on observations he made during his own training with Denny-Brown in Boston and later when he was a junior faculty member at Columbia while Merritt was chief there. He wrote that Denny-Brown

> spoke fondly of Merritt. He once commented Merritt was most considerate in continuing as Acting Director of the Neurological Unit of the Boston City Hospital during the two long periods in World War II when Denny-Brown served in the British army. During the times that Denny-Brown could return to Boston, the two of them

frequently drove home together from the hospital, and he commented with a grin that Dr. Merritt often suggested that they stop along the way so that he could have an ice cream soda before going home for dinner. Much of the presumed animosity between these two has been fictions created by others, not the two of them.[13]

Whatever their differences, Denny-Brown and Merritt both made major contributions to the elucidation and treatment of neurological diseases and trained another generation of neurology leaders. Both are ensconced in the pantheon of "great neurologists."

Notes

1. Braceland FJ. Cited by Braceland and applied to H. Houston Merritt in memorial service for Dr. Merritt, 1980. Archives and Special Collections, Long Health Science Library, Columbia University.
2. Cecelski DS, Tyson TB, eds. Democracy betrayed. The Wilmington race riot of 1898 and its legacy. Chapel Hill: University of North Carolina Press, 1998.
3. Mrs. Dessie C. Merritt [obituary]. Wilmington Morning Star, January 9, 1946.
4. Wilmington boy wins high honor at N.C. university; H. H. Merritt, Jr. among eight making highest possible grade. Wilmington Dispatch, January 25, 1920.
5. Will leave today for the far west. Wilmington Star, January 8, 1924.
6. Houston H. Merritt [obituary]. Wilmington Morning Star, May 8, 1945.
7. White BV. Stanley Cobb: a builder of the modern neurosciences. Boston: Francis A. Countway Library of Medicine, 1984, p 164.
8. Friedlander WJ. Putnam, Merritt and the discovery of Dilantin. Epilepsia 1986;27(suppl 3):S1–S21; comments S7.
8a. White BV. Stanley Cobb: a builder of the modern neurosciences.
9. Telephone conversation with author, May 5, 2006. Dr. Foley is a proud Irish-American Catholic, a renowned neurological alumnus of the Boston City Hospital, and the long-time Head of Neurology at the Metropolitan Hospital in Cleveland.
9a. White, Stanley Cobb, 164.
10. Personal communication with Joseph Foley, 2007.
11. Gilman S, Vilensky JA. How Denny-Brown came to Harvard. Harvard Medical Alumni Bulletin 1997(Winter):46–50.
11a. Personal communication with Joseph Foley, 2007.
11b. S. Gilman E-mail to the author, April 29, 2008.
12. J. S. Meyer to the author, May 10, 2006.
12a. Ibid.
13. S. Gilman E-mail to the author, April 29, 2008.

11

Merritt Moves to New York: Montefiore and Columbia— Another Tale of Two Hospitals

Montefiore Hospital sits on Gun Hill Road and 210th Street in an unfashionable area of the Bronx, a borough of New York City. The hospital opened there in 1913, but it originated in Manhattan as the Montefiore Home for Chronic Invalids in 1884.

The physician who transformed the institution from one restricted to chronic care to a modern all-purpose hospital was Ernst Boas, son of Franz Boas, the famed anthropologist. Ernst himself was a cardiologist, ardent proponent of national health insurance for all, founder of Physician's Forum, and a straight-talker. About Charles Elsberg, one the most famous neurosurgeons of the day, Boas said, "Elsberg especially causes many difficulties. There is nothing we can do with him and I am afraid we shall have to get rid of him."[1] Elsberg was also chief of Neurosurgery at the Neurological Institute at about the same time. Boas also said that his famed chief of Medicine, B. S. Oppenheimer, "has no push and will accomplish nothing by himself."[1a] Boas became head of the hospital in 1921; he himself was 31 when he battled the giants who were to work for or with him.

Boas started a house staff training program, organized the staff into appropriate departments, recognized the need for patients with acute—not chronic—conditions to attract attending physicians, and fostered the incorporation of research into patient care. David Marine, one of the first clinical investigators in the United States, was on the staff at Montefiore. Marine identified the importance of iodine deficiency in causing "goiter," a cystic swelling of the thyroid gland; his observation led to the prevention of goiter by adding iodine to table salt, a now universal practice. Marine was paid a salary, not a patient-dependent fee-for-service; this was an unusual financial arrangement at the time (and one opposed by the American Medical Association), but it established another tradition at Montefiore, fostering the combination of research with clinical practice.

One of Boas's achievements was the formal establishment of a department of neurology, doing so in affiliation with Columbia University. In 1928, Harvard, Pennsylvania, and Montefiore-Columbia were the only neurology departments in the country.[2] The hospital was then in transition to acute care, but some patients were kept for years—without alternative possibilities but also for teaching medical students and house staff. In time, some patients were used as subjects for the specialty examinations in neurology and psychiatry. The candidate would have to take a medical history and carry out the neurological examination under the critical gaze of an experienced neurologist. After years of being subjects again and again, some patients learned what findings on examination were important for diagnosis.

One patient was asked by the test supervisor to comment about a candidate who had just finished the examination. "Well," the patient said in speech characteristic of the altered tempo and rhythm signifying disease of the cerebellum, "he found the nigh-stag-mus, and he recognized my cer-e-bell-ar speech. He even found the Ba-bin-ski signs." But, he continued, pointing to the top of his head, "He *missed* the eu-*phor*-ia!"[3]

In 1928, Boas resigned and Ephraim M. Bluestone was appointed director, the first to hold that position without a simultaneous clinical practice. Bluestone also favored the full-time salary system.

Henry L. Moses, a wealthy lawyer, became chair of the board in 1936. Moses's law firm represented major industrial firms in the United States and foreign governments. He was a doer and joined Bluestone in two goals, expanding the physical plant at Montefiore and finding outstanding chiefs of service.

In the 1930s, many of the staff neurologists at Montefiore had simultaneous appointments at Mt. Sinai Hospital. It was still difficult for Jewish physicians to find a place on the staff of other hospitals in the city. David Marine was an isolated Protestant physician in a Jewish hospital.

Among the luminaries were Moses Keschner, Moses Madonic, Abraham Rabiner, and Israel Wechsler. Neuropathology started with Charles Davison, who had been born in Romania in 1891, served as neuropathologist at Montefiore from 1932 to 1953,[4] was president of the American Association of Neuropathologists in 1948, and ended his career as a psychoanalyst.

The hospital itself gained a reputation as socially innovative, one of the first to have a social service department, a home care program, and a prepaid group medical practice. (The key words in that third achievement were "prepaid" and "group." In the United States, it has not yet been possible to achieve these features in a national health insurance program.)

As World War II was drawing to a close, Bluestone expressed the view that "mental disease is as much entitled to the interest of the modern hospital as any other physical disease"[4a] and announced plans to build a new 200-bed institute of neuropsychiatry.

In 1944, Bluestone and Moses approached Merritt. How that came about is not recorded, but it is likely that Harry Zimmerman had a hand in it. He and Merritt were good friends, having been interns in medicine together on the Yale service at New Haven Hospital. Zimmerman was now returning from service in the Navy on the island of Guam, and he had had enough of Milton Winternitz, his tough-guy dean at Yale. Montefiore was a suitable opportunity. Merritt's subservient position to Denny-Brown certainly made him ripe for an attractive offer to move. Merritt went to Montefiore with two of his Boston City Hospital house officers, Sidney Carter and Daniel Sciarra. Both of them claimed to have been Merritt's last neurology resident and Denny-Brown's first resident at the Boston City Hospital. At that time, Carter had no notion that he would eventually become a pediatric neurologist at the Neurological Institute. Sciarra was a popular clinician and a devoted teacher.

Within a few years, the Merritt team had established an active clinical service. Adding to the original staff members, Arnold Friedman set up the Headache Unit, the first of its kind in the United States; he was director of that unit for 30 years, retiring to Arizona.[5] Melvin D. Yahr came to Montefiore as a clinical fellow and was Merritt's close associate for three decades before he became chief of Neurology at Mount Sinai Hospital.

Someone asked Merritt how he was getting along as the only Christian on the professional staff of a Jewish hospital. Merritt replied: "I get along with them very well. But they don't seem to get along with each other."[5a]

Merritt was as popular as ever, but Putnam was already in deep trouble at the Neurological Institute and left in 1946. The search committee for his successor turned to Merritt, and he moved again in 1948, this time to the Neurological Institute. With him went Carter, Sciarra, and Yahr. Friedman also had an appointment at Columbia and, after the others had moved, continued to teach at both Montefiore and the Neurological Institute.

When they arrived at the Neurological Institute, Merritt's group found a hospital with a sharp division between "private patients," who paid a fee to an attending neurologist for personal attention, and "ward patients," who had no insurance for doctors' fees and whose hospital bills were covered by city and federal programs. Care of ward patients was the responsibility of house staff residents in training under the supervision of an attending neurologist. Teaching was restricted to the ward patients, but the interactions and cross-checking by house officers, attending physicians, and pair and group discussions in teaching sessions made the care of these individual patients the best possible.

Before long, private patients were included in the teaching programs, assuring that they too would also get optimal care. Fears that patients would resist teaching functions vanished as the patients recognized the benefits. The emphasis on teaching made positions on the house staff a prize, and more positions became available when the National Institutes of Health (NIH) began to support training programs. Under Merritt's direction, the neurology training program became one of the most desirable in the country. The neurosurgery program under J. Lawrence Pool gained similar status.

One problem was a unit for children, which had been established by the legendary Bernard Sachs (of Tay-Sachs disease), who had moved from the Neurological Institute to found the Department of Neurology at Mt. Sinai Hospital. He was followed at the Neurological Institute by the avuncular Louis Casamajor, but neurology house officers came to the Neurological Institute with training and experience in treating adults, not children. When Casamajor retired, the beds for children were moved to the Babies Hospital. That, however, left a gap. Pediatricians could care for the children, but who would do the neurology?

One day, according to legend, Merritt called Sidney Carter into his office and asked him to take charge of child neurology. "But I am not a pediatrician," Carter responded. Merritt continued, "No neurologist here is a pediatrician and the pediatricians do not know any neurology. It will be easier for you to learn what pediatrics you need than for a pediatrician to go through neurology training."

Carter was not the first pediatric neurologist. Douglas Buchanan was already working at the University of Chicago, David Clark was recognized at Johns Hopkins, and Philip Dodge was at the Massachusetts General Hospital. However, Carter was the one who developed the first NIH-supported training program in child neurology. He attracted a superb coterie of house officers, who then fanned out to head new programs throughout the country.

According to one widely-believed story, Merritt once said that "research is the frills on the panties of patient care," but he fully supported the quartet of scientists appointed by his predecessor, Tracy Putnam—Nachmansohn, Kabat, Mettler, and Grundfest. Indeed, he had written a book on clinical neuroanatomy with Mettler and Putnam. And soon after they arrived at the Neurological Institute, Merritt had Yahr doing research with Kabat on cerebrospinal fluid. He also supported Gilbert Glaser, who was already on the staff at the Neurological Institute and joined Merritt in a trial of adrenal steroid therapy in neurological diseases. Glaser left to become the first chief of Neurology ever appointed at Yale. A division first, neurology there later became a full department under Glaser's leadership.

In the mid-50s, the NIH awarded Merritt a grant to set up the Clinical Research Center at the Neurological Institute. The renovations for laboratories at the Neurological Institute were completed in 1962, led by Robert Fishman and myself. Now, a new self-renewing system was in place. Neurological Institute residents would join the research labs for training and would soon be eligible for NIH grants themselves.

In the meantime, Merritt became dean of the medical school in 1958, a position he held until 1970. Simultaneously, he was a university vice president, responsible for administration of the School of

Public Health, the School of Dentistry, and the School of Nursing. As dean, he obtained a donation of $5 million from William Black, who had created the Chock Full o'Nuts company and was interested in parkinsonism. Black, a no-nonsense executive, was distressed to find that a close friend had Parkinson disease and that no one was doing any research. So he founded the Parkinson's Disease Foundation in 1957. Merritt was president of the foundation and Black was chair of the board of directors. Melvin D. Yahr, Merritt's most trusted academic partner, became scientific director of the foundation and was largely responsible for recruiting research scientists for the new 20-story research building. By the Deed of Gift from William Black, one floor of the new building was reserved for Parkinson disease research. Yahr put together a formidable group of investigators, including Erminio Costa, a molecular pharmacologist; Virginia Tennyson, an electron microscopist; and Roger Duvoisin, a neurologist and former student of Merritt's who later became chair of Neurology at the Robert Wood Johnson School of Medicine in New Jersey. This multidisciplinary approach was a unique organization for the time but is now commonplace, and Yahr was lauded for his leadership.

Among the first Neurology residents who stayed for postdoctoral research training were Stanley Fahn, Robert Layzer, Audrey Penn, Donald Schotland, Joseph Willner, Frank Yatsu, and Neil Raskin.

Neuropathology was another source of research strength with Abner Wolf and David Cowen. Paul F. A. Hoefer and Eli S. Goldensohn combined with the group from Montefiore to provide a strong program in epilepsy. Neurosurgery was directed by J. Lawrence Pool, a pioneer in operations on blood vessels of the brain, especially anomalies called "aneurysms" (bubbles on arteries that are formed by the weakened cell walls of the vessels). The weak arterial walls would sometimes rupture and result in hemorrhages into the substance of the brain or the spaces around the brain, with often disastrous results.

Merritt's training program attracted outstanding physicians from the start. Some remained on the attending staff at the Neurological Institute; at least 45 of them became chiefs of neurology at other medical schools.[6]

One outstanding department was based on a team of Merritt's students led by Saul R. Korey, who had been chief resident at the Neurological Institute in 1948, before Merritt arrived there. Korey then learned biochemistry with Severo Ochoa, a Nobel laureate at

New York University, and returned to Columbia to work with neuro-chemist David Nachmansohn. He spent most of his time in the laboratory but also made teaching rounds with patients. He and Merritt had enough personal contact to share mutual admiration.

In 1952, Korey became an associate professor at Western Reserve University School of Medicine in Cleveland, returning to New York three years later to become the first chair of Neurology at the new medical school at the Albert Einstein College of Medicine. There, he recruited other Neurological Institute alumni who had also been trained by Merritt: Robert Katzman, Labe C. Scheinberg, Elliott Weitzman, and Isabelle Rapin. He added neuropathologist Robert Terry and neurochemist Kuni Suzuki. Immediately, Korey reordered clinical research for the United States. He organized a team approach that focused on a single problem, what would now be called "multi-disciplinary research" involving clinicians, biochemists, physiologists, and electron microscopists. Bringing together investigators with different approaches to focus on one or similar diseases is now called a "program project" or a "clinical research center." Korey's team worked on Tay-Sachs disease and on Alzheimer disease, which became a long-range research project for the rest of Katzman's distinguished career at Montefiore and then as head of Neurology at the University of California, San Diego. Korey's group became a transforming model for other institutions to follow. Off to a great start, Korey tragically died in 1963 at age 45 of pancreatic cancer. His legacy lives on in the continuing power of patient-oriented research at a fundamental or "molecular" level.

Merritt was proud of his trainees at Einstein, and he had reason to be proud of his work at Columbia. During those years Merritt showed himself to be a canny leader, melding the work of disparate talents on his faculty. He had a unique sense of humor, which he wielded as a benign weapon in dealing with people. His aphorisms were legendary:

1. "The first rule of public speaking, essential for all trainees headed for an academic career: Empty your bladder."
2. "There is no such thing as tainted money, only it 'taint enough."
3. "Never turn down a job you haven't been offered."
4. When he was dean, he said, "I have two categories of faculty. The part-time people are away part of the time. The full-time people are away all the time."

5. "My faculty is like the Strategic Air Command. Half of them are in the air at any given moment."

6. "A good dean is one who can say 'no' and make you think he said 'yes.'"

7. "An assistant dean is a mouse practicing to be a rat."

8. "A mugwump is a person who sits on the fence, with his mug on one side and his wump on the other."

9. "Medical specialties originated about the body's orifices—ophthalmology for the eyes; otolaryngology for ears nose and throat; gynecology; proctology; gastroenterology—all of them. When they were finished, they invented psychiatry to deal with the body as a whole."

10. On the universal cure, prednisone: "It is good for coughs, colds, sore holes, and pimples on the belly."

11. In his last days, one of the symptoms of normal pressure hydrocephalus was loss of bladder control, which he described candidly: "When I was young, I had trouble remembering to pull up my zipper. Now, it is the other way around."

Sometimes his humor was more subtle. One story was related by John Romano, one of Merritt's house officers at the Boston City Hospital and later a leading professor of psychiatry at the University of Rochester:

Almost every day, three or four of us would accompany Merritt to Goldstein's Delicatessen, a basement snack store, with an elaborately ornate art deco jukebox. Regardless of which button you pushed, "The Beer Barrel Polka" exploded into the air. Coins would be tossed to decide who would pay for the Cokes and also who would feed the jukebox. Invariably, Merritt won; that is, he never paid. Many years later, in 1976, I wrote to him and said, "And perhaps now that you have prestige and security and the high esteem of your friends—perhaps now you can confess and show to the world the two-headed coin that you used so skillfully in winning free Cokes from us at Goldstein's. I am told that repentance is a good thing.[7]

One week later, Merritt responded, "I can assure you that I never had a two-headed coin, and if I was fortunate in winning a few Cokes from the boys at Goldstein's, it was probably the just reward for my honest and exemplary conduct."

Could there have been a dark side? I knew him for 29 years and never heard him utter an ethnic slur. Robert Fishman, however, recalled a 1966 conversation when he left Columbia to become chair at the University of California in San Francisco. Merritt advised him, "Don't make the same mistake I did. Don't appoint too many Jews." To me, that remark is out of character because so many of Merritt's princes were Jewish and he was a close friend to several.

Another story was told by some medical students and is recorded in a report about the state of women and minority students at the Columbia medical school.[8] The 1979 report included an episode that unflatteringly portrayed a famous but unnamed neurologist who had been dean; it could only have been Houston Merritt. The authors described an episode in which the Merritt doppelganger persistently asked medical students if they knew which "race" is most prone to alcohol addiction, prodding them for the answer he meant to be "African American."

That attribution, however, was also out of character because Merritt's considerable experience with the neurological complications of alcoholism was based on the Boston City Hospital, where the poor and the alcohol abusers were predominantly Irish, not blacks.

One of Merritt's great achievements as dean was the forging of an affiliation of Harlem Hospital and Columbia University, an enduring partnership that brought the best of academic medicine to an inner-city hospital. Reciprocally, the affiliation has brought the medicine of poverty to the lives of Columbia medical students. During the dark years of McCarthyism, Merritt was "steadfast in supporting members of his department who were then under attack for divergent political views."[9]

Merritt's own research achievements resulted in the publication of books on the cerebrospinal fluid, epilepsy, neurosyphilis, headache, stroke, multiple sclerosis, and genetic diseases. In these books he recorded studies on human patients, not animal models. In the process, he was one of the founders of what is now called "clinical investigation." He also recognized the importance of working with advocacy groups devoted to individual diseases. For instance, both he and Tracy Putnam testified in 1948 on the need for a national neurological institute at the NIH. And they were both active on the executive boards of the National Multiple Sclerosis Society (of which Putnam was the first chair) and the Epilepsy Foundation.

Merritt also wrote an enduring textbook of neurology. The sixth edition was published two weeks after Merritt's death. In 2005, the eleventh edition appeared, with chapters written by his former students. The twelfth edition is expected in time for the celebration of the centennial of the New York Neurological Institute in 2009.

His clinical skills brought him international fame, and celebrities called him for consultation, including Dimitri Shostakovich, Portuguese dictator Antonio Salazar, U.S. president Dwight D. Eisenhower, and actress Rita Hayworth. When he was at Montefiore Hospital, the chair of the Board of Trustees was Henry Moses, who had a severe stroke. Merritt became so involved in caring for Moses that he became part of their family. As one result, Lucy Moses became a lifelong benefactor of both Montefiore Hospital and the Department of Neurology at Columbia.

According to an obituary for Merritt,

> His efficiency and capacity for work were extraordinary, perhaps best illustrated by an anecdote told by William Amols [a former resident of Merritt's], who once visited the Merritts at their summer home. Mrs. Merritt met Amols at the door and told him that Dr. Merritt was at the waterfront. As Amols approached, he saw Dr. Merritt from the rear and, as he drew nearer he saw that Houston was "relaxing." He was sitting in a beach chair, facing the water. On his left was one radio, announcing the Giants game. On the right, a second radio was tuned to the Yankee game. In the middle, the Professor was doing the crossword puzzle—in ink, as usual.[9a]

Merritt also had a reputation as a fundraiser, a down-home codger who somehow could wangle donations from wealthy patrons. Most of these encounters were simple and direct. One involved his wily relations with Huntington Hartford, heir to the fortune of the family that created the Atlantic and Pacific (A&P) stores, one of the first national chains of supermarkets. Hartford believed in graphology, the science of handwriting, as a key to the "total personality," which is "made up not only of the date of birth but of heredity, environment and daily moods."[10] Hartford published a book on the subject and asked Merritt to write the foreword, a challenge Merritt accepted, understanding the fine line between implausibility and doing a favor for a friend (and, undoubtedly, a possible contribution to the Department of Neurology). He explained how the physiological activity of the brain controls the finger movements of handwriting: "The neurologist can understand how injury to the basal ganglia in Parkinson's disease influences the size and shape of the letters and the flow of the script as the result of rigidity and tremor of the muscles." He added that "the physician is not aware of the importance of such factors as rhythm, pressure, beginning strokes and the like, either because he is not trained to recognize them or because he considers them to be too erudite to be evaluated accurately."[10a]

The exchange was typical of Merritt. He did not believe in graphology, but he explained his views politely and respectfully. Hartford responded grandly by donating a Rembrandt portrait to Columbia. Until the student antiestablishment revolt of 1968, the painting had been displayed in the office of the university president, Grayson Kirk. On until Wednesday, April 24, 1968, undergraduates occupied the office, so the painting was taken down and packed away for safekeeping.[11] It remained in hiding until it was finally sold in 1975 for $1.75 million, enough for two endowed chairs, one for Psychiatry and one for Neurology.

Notes

1. Dorothy Levenson. Montefiore. The hospital as social instrument, 1884–1984. New York: Farrar, Straus & Giroux, 1984, p 117.
1a. Ibid.
2. Ibid., 127.
3. Ibid., 128; also, author's experience with the same patient.
4. Brenner C. Charles Davison, 1891–1965. Psychoanal Q 1966;35:275–276.
4a. Dorothy Levenson. Montefiore. The hospital as social instrument, 1884–1984.
5. Dr. Arnold P. Friedman, 81, dies; authority on migraine headaches [obituary]. New York Times, September 20, 1990.5a. Another apocryphal story
6. From the Boston City Hospital: Raymond D. Adams, Robert Aird, Charles D. Aring, Richard Chambers, Thomas Farmer, Francis Forster, W. Jordan, Augustus C. Rose, Ephraim Roseman, Milton Rosenbaum, T. Von Storch. From the Neurological Institute: Gary Abrams (Neurology and Rehabili-

tation Medicine, University of California, San Francisco), Milton D. Alter (Temple), Stanley H. Appel (Baylor), James Austin (Colorado), Susan Bressman (Beth Israel Hospital, New York), Rosalie Burns (Women's Medical School, Philadelphia), Peter Dunne (University of Florida, Tampa), Robert A. Fishman (University of California, San Francisco), Sid Gilman (University of Michigan), Gilbert Glaser (Yale), Melvin Greer (University of Florida, Gainesville), Warren Grover (Child Neurology, Temple University), Donald H. Harter (Northwestern University), T. R. Johns (Virginia University), B. Jubelt (University of Syracuse), R. Katzman (Albert Einstein College of Medicine, University of California, San Diego), Charles Kennedy (Children's Hospital of Philadelphia), Edward Gildea (chair of Psychiatry, Washington University, St. Louis, 1942–1962), S. Korey (Albert Einstein College of Medicine), J. Mendell (Ohio State), J. P. Mohr (University of Alabama, Mobile), Frank Morrell (Stanford University), Theodore L. Munsat (Tufts University), Jack Pellock (University of South Carolina), L. D. Prockop (University of South Florida), Clark D. Randt (Western Reserve, New York University), L. P. Rowland (Pennsylvania, Columbia), L. C. Scheinberg (Albert Einstein College of Medicine), James Schwartz (Emory University, Child Neurology), W. A. Sibley (University of Arizona), L. Weisberg (Tulane University), E. Weitzman (Montefiore Hospital, New York), Robert Williams (Psychiatry, University of Texas, Southwestern), Melvin D. Yahr (Mount Sinai School of Medicine, New York), F. Yatsu (University of Oregon; University of Texas, Houston), D. K. Ziegler (University of Kansas), E. A. Zimmerman (Albany Medical College).

7. Romano J. Harvard Neurology, Boston City Hospital, and Peter Bent Brigham Hospital 1938–1941. Arch Neurol 1998;45:1263–1266.

8. "The double image. Women and minority students at P&S" was prepared by Mary Roman and Nancy Anderson, both in the medical school class of 1980, and submitted in October 1977. For "initiating this project," the authors thanked Esther Rowland, who was then associate dean of studies at Barnard College and advisor to the two authors of the report. After first being received by two medical school deans, the report was referred to the Medical Faculty Council; receipt was acknowledged in April 1979 by Michael Gershon, MD, chair of the Department of Anatomy, and Audrey S. Penn, MD, chair of the Executive Committee and professor of neurology; but no action was ever taken. The source of the comments about Merritt was Leslie Davidson, MD, MSc, at the time a classmate of Roman and Anderson and in 2007 a professor of public health at the Columbia School of Public Health. In February 2007, she said in a conversation with the author that she was the unnamed source of the quotation (on pp 25–26 of the 62-page report) and gave permission to be quoted. The report has been deposited in the Special Collections Division of the Health Science Library of Columbia University.

9. Rowland LP. H. Houston Merritt 1902–1979. Neurology 1979;29:277–279.

9a. Rowland LP. H. Houston Merritt 1902–1979.10. Hartford H. You are what you write. New York: Collier Books, 1973, p 3.

10. Hartford H. You are what you write. New York: Collier Books, 1973, p 3.

10a. Ibid.
 11. McCaughey RA. Stand Columbia. New York: Columbia University Press, 2003, p 444.

12

Putnam in California

Charles Carton was a medical student at Columbia and took care of some of Tracy Putnam's patients then. Carton became a neurosurgical resident a few years later and ultimately became chief of Neurosurgery at the Cedars-Sinai Medical Center in Los Angeles, where he again encountered Putnam.

Although he said he was "just a student," Carton assisted Putnam in operations and followed the patients after surgery. He recalled Putnam as "a pretty cold person" and "very precise, a little prissy."[1] Carton remembered that Putnam's experimental surgery to relieve involuntary tremors was of limited value, the patients were not "doing very well," and they were therefore assigned to a "secret service."

Carton started his neurology residency before he became a neurosurgical trainee in 1947, so he had a youthful but memorable impression of Putnam's problems as chief. He thought the senior neurosurgeons were unhappy with Putnam because he was doing experimental surgery that did not always have a good outcome and they did not consider him a very good surgeon. Carton added another "fault": Putnam had not been trained at the Neurological Institute and was therefore an "outsider."

Putnam arrived in Los Angeles in 1947, after two traumatic years of battle in New York. He landed another highly responsible position as chief of Neurosurgery at the Cedars of Lebanon Hospital, in private practice. Carton was also a historian of that hospital. He recorded the number of operations each surgeon did; Putnam's total for the first six months that year was a paltry six, but about the same as more established surgeons there. Whether he was more active in later years is not known. A decade after he arrived, he was no longer chief-of-service but still held an appointment.

During those years, Putnam continued some of his earlier research on the surgical treatment of the involuntary movements of parkinsonism as well as disorders of the fluid within cavities of the brain; one condition was enlargement of these spaces, or "hydrocephalus," which means "water on the brain." He also tried to use radioisotopes to find brain tumors. His early interest in the neurological basis of vision[2] and his lively imagination were on display when he tried to restore vision to three young adults who had been blind as a result of injury to the part of the brain that serves vision and who had atrophy of the optic nerves in the eyes. Putnam inserted electric wires through the skin and skull into the posterior part of the brain, the calcarine fissure of the occipital lobe, where vision is integrated.[3] Stimulating the wires electrically or with a photoelectric cell that converted light to electricity (an "electric eye") activated neurons in the brain, which was perceived by the subject as a light flash. The evidence proved that the visual cortex could still function despite years of inactivity. That experiment was the forerunner of modern experiments in which a tiny video camera is mounted on the eyeglasses of a blind person to power electrodes in the same part of the brain and thereby restore limited but useful vision.[4,5]

Putnam's later years were, however, dominated professionally by multiple sclerosis. He had a private practice and was considered an expert on that disease. His treatment was based on his personal theory that the nerve damage in that disease resulted from abnormal clotting that obstructed veins and somehow injured the brain and spinal cord. He therefore prescribed dicumarol, an early blood thinner that has since been replaced in the treatment of heart disease and stroke by warfarin (Coumadin), which is less likely to become overactive and cause unwanted bleeding. In fact, Putnam's theory never gained widespread acceptance, which became evident in an analysis of practice patterns in Los Angeles.

Colin L. Talley,[6] a medical historian, compared Putnam's treatment in private practice with that of the Neurology Clinic at the University of California, Los Angeles (UCLA), during the years 1947–1965, a time when there was no effective therapy for the disease itself. And there were no randomized clinical trials of treatment, a practice that did not appear until the 1960s and is now the basis of "evidence-based medicine." A modern clinical therapeutic trial is based on the comparison of a large number of patients in at least two groups treated identically, except that one group is given the active agent or drug and the other an inactive placebo. Neither the patient nor the treating physician knows the status of an individual patient. In contrast, either the UCLA physicians or Putnam would have given patients a treatment and merely observed what happened. Putnam saw his last new patient with multiple sclerosis in 1965. He was still prescribing anticoagulation in 1970 when he was 76 years old.

At UCLA, the basis of therapy was a diet low in fat and higher in supplemental nutrition. They gave more anti-inflammatory drugs, adrenocorticotropin(ACTH), and cortisone than Putnam did. They also gave more sedatives, painkillers, and muscle relaxants. The neurologists at UCLA were more likely to be the patients' primary physicians for continuing care. Putnam, in contrast, was likely to prescribe anticoagulants.

According to Talley's analysis of Putnam's case records, the neurologist prescribed that drug to 74 patients but followed only 17 closely. For those 17, Talley found "only equivocal success" and Putnam actually did not see many multiple sclerosis patients: about one a week in 1948 and two to four a month between 1947 and 1961. "He did not see a sufficient number of patients at regular enough intervals for the negative evidence to accrue to him. Because of this, his belief in the efficacy of the treatment could withstand what to a later analyst seemed clearly equivocal or contradictory evidence."[7]

Within a few years, Putnam's drawing power shrank and fewer and fewer patients came from areas outside Los Angeles. Nevertheless, in 1947, Putnam became chair of the Medical Advisory Board of the National Multiple Sclerosis Society. That year, the society sent out an advisory message to 70,000 physicians, which recommended "no specific treatment" and did not mention anticoagulation.[8] The low-fat diet used at UCLA also disappeared from view as new and more effective treatments came on the scene.

As time went on, Putnam apparently had financial difficulties. Starting in 1947, he was designated chief of Neurosurgery at the Cedars of Lebanon Hospital; but he was in private practice, and his term ended in 1958 under circumstances that are not documented. He continued to serve on the staff of the hospital until he died in 1975. Putnam's problems led financier Jack Dreyfus to contribute a large sum for Putnam. Carton called it the "Dreyfus Medical Foundation

Award" for 1969 and stated the amount was $75,000. There had never been such an award before or after and, although the prized achievement was Dilantin, Merritt was not included. The award was given at a dinner in Boston, and by the year 2000, the amount would have approximated much more, probably more than a million dollars.

Another possible financial indicator was Putnam's career as a film actor. He played the role of a physician in a film called *The Slime People*. According to the movie's website, Putnam's name is not mentioned in advertisements for the 1962 original, but there he is in the cast of characters for the 2001 DVD version.

Tracy Putnam's son, Jock, was a sound engineer for *Slime* and for many films that were equally negligible; he may have arranged the hiring of his father. On the other hand, Putnam senior had a theatrical bent and was especially known for suddenly appearing at departmental affairs in the guise of some foreign dignitary with an accent.

His old partner, Houston Merritt, maintained an interest in epilepsy throughout his life; and under Merritt, the Neurological Institute of New York became a center for research and treatment of the condition, persisting even after Merritt left the scene. Putnam's interests shifted to multiple sclerosis and surgery for movement disorders. His dwindling contacts with people who had seizure disorders may have led him into the erroneous belief that antisocial acts of violence could be manifestations of epileptic activity in the brain and when he wrote that both Lee Harvey Oswald, the assassin of John F. Kennedy, and Jack Ruby, the killer of Oswald, could have had epilepsy.[9] In fact, at Ruby's trial, Putnam's other Boston colleague and pioneer in brain wave diagnosis, Fred Gibbs, gave testimony in the same vein.[9a] However, that testimony was demolished by Frank Forster, who said that such planned acts of violence could not be epileptic.[10] Gibbs claimed that the judge told him "What I want to know and also what the jury wants to know is whether Ruby could tell right from wrong." Gibbs responded, "No, the EEG wouldn't show that," and the judge replied, "You are dismissed."[11]

Notes

1. Charles Carton, MD, telephone interview with the author, October 23, 2003.
2. Putnam TJ, Putnam IK. Studies on central visual system: anatomic projection of retinal quadrants on striate cortex of rabbit. Arch Neurol Psychiatry 1926;16:1–20.

3. Button J, Putnam T. Visual responses to cortical stimulation in the blind. J Iowa Med Soc 1962;52:17–21.
4. Dobelle WH. Artificial vision for the blind by connecting a television camera to the visual cortex. ASAIO J 2000;46(1):3–9.
5. Fernandez E, Pelayo F, Romero S, et al. Development of a cortical visual neuroprosthesis for the blind; the relevance of neuroplasticity. Neural Eng 2005;2:R1-R12.
6. Talley C. The treatment of multiple sclerosis in Los Angeles and the United States, 1947–1960. Bull Hist Med 2003;77:874–899.
7. Ibid., 885–886.
8. Ibid., 888.
9. Putnam TJ. The adolescent misfit. West Med J 1964;126:115.
9a. Ibid.
10. Gutmann L. Jack Ruby. Neurology 2007;68:707–708.
11. Hughes JR, Stone JL. The Gibbs years: from Harvard in the past to Illinois in the present. Clin Electroencephalogr 1990;21:183–187.

13

Merritt's Last Days

Houston Merritt was famous among neurologists for many achievements; none was respected more than his monograph on the cerebrospinal fluid (CSF). A disorder of CSF circulation may have killed him indirectly, and the irony was compounded because he had long doubted the validity of that particular diagnosis, which is called "normal pressure hydrocephalus" (NPH).

Merritt's doubts were based on his skepticism toward the theory about the cause of the condition and because, although there is supposed to be an effective treatment, it is still difficult to predict which patients will improve.

The syndrome is characterized by a triad of symptoms: unsteady gait, dementia, and urinary incontinence. The brain damage is attributed to problems in the circulation of the CSF. The fluid normally cushions the brain against physical injury; when the head moves, the brain does not bang against the hard bones of the skull because it floats in the fluid.

The fluid arises from blood vessels in a chamber within the brain and then flows through more deep caverns called "ventricles" until it reaches a series of microscopic holes in the membranes that allow it to reach the external surface of the brain and continue flowing.

When the moving CSF reaches the external surface on the top of the brain, specialized blood vessels allow entry back into the blood. If scar tissue occludes passageways between the ventricles or if a hemorrhage into the CSF or pus from an attack of meningitis interferes with the absorbing channels, the flow is dammed back, the ventricles enlarge or "blow up," and pressure in the system increases to danger-ous levels. This condition is called "hydrocephalus," or "water on the brain," and the hazard is the pressure, not the amount of water. If the process is slow, the ventricles may enlarge without a concomitant in-crease in pressure or the brain tissue may adjust to the high pressure and so reduce the pressure; how that happens is the mystery of NPH.

The treatment of NPH is to insert a tube that drains CSF from the ventricles or spaces around the spinal cord and diverts it into the ab-domen or directly into a large blood vessel. Restoring the circulation somehow can reverse the clinical disorder but not in all patients, and it has been difficult to identify which patients will respond favorably.

Merritt's problem came in 1971, when he was 69 years old and was first treated for high blood pressure. In 1975, he abruptly had weakness of the left arm and leg for 24 hours, presumably a small stroke. In 1977, he fainted and irregular heart action was found. He was treated with digoxin and had no further episodes, but high blood pressure and arteriosclerosis were held responsible for the cardiac symptoms.

In about 1973, Merritt's friends noticed that his gait was slow and unsteady. His posture became stooped. Some experts thought he had Parkinson disease and prescribed levodopa in 1975 without effect. The problem became marked in 1977, and he often felt as though he were being pushed forward when he went down stairs or down an in-cline, a parkinsonian feature called "propulsion." Once, he was found leaning against a wall because his balance was so impaired. In exami-nation by Dr. C. T. Vicale, then the senior clinician at the Neurological Institute, Merritt had what are called "upper motor neuron signs," indicating dysfunction on both sides of the brain; these findings also implied more extensive brain damage than expected in Parkinson disease. Computerized tomography (CT) showed evidence of slight focal brain atrophy as well as enlarged ventricles in a pattern suggest-ing NPH, according to Dr. Sadek Hilal, chief of Neuroradiology. An isotope study of the CSF circulation was analyzed by the head neuro-surgeon, Edward B. Schlesinger; abnormally, isotope injected into the CSF by spinal tap failed to appear over the cerebral hemispheres. That failure was attributed to an impaired CSF circulation, as might be ex-pected in NPH. The cellular and biochemical contents of the fluid were normal, which was also compatible with that diagnosis.

The symptoms cleared in a few days, and he continued his daily activities on committees, actively revised his textbook of neurology, and completed the crossword puzzle daily. His associates noted no intellectual decline.

However, the gait disorder progressed, and in September 1978 he was using a wheelchair at home. The parkinsonian signs led to treatment with amantadine and trihexyphenidyl (Artane) on the recommendation of Dr. Melvin Yahr, chief of Neurology at Mt. Sinai Hospital and a close friend of Merritt's. In October, Merritt had another 48-hour episode of right-sided weakness of the arm and leg. Emotional lability, another upper motor neuron symptom, was manifest by episodic and inappropriate crying. He was admitted to the hospital on November 5, 1978. Tendon reflexes were more active on the left, residual effects of the last minor stroke. He had no urinary problems or clear signs of parkinsonism. The major finding on examination was the slow and unsteady gait.

CT showed no change from earlier studies; that is, the ventricles were no larger, even though the symptoms were worse. The cisternogram again showed no isotope on the convexity of the brain, but Schlesinger considered the pattern atypical and did not advise shunting. "Therapeutic" spinal taps were done, with sharp drops in CSF pressure after small amounts of fluid were withdrawn. This was found on November 6, 10, and 11. Atypically, the protein content of the fluid was excessively high. Normally less than 45 mg, values on the three separate days were 76, 83, and 173. This abnormality was not a recognized feature of NPH but would be consistent with diffuse vascular disease of the brain. Nevertheless, his gait improved unequivocally for three days and then returned to an impaired state. On November 12, he was discharged with three diagnoses: cerebrovascular disease, hypertensive vascular disease, and "possible" NPH.

On December 4, Dr. Stanley Fahn wrote that Merritt's cognition was better than prior to admission, as was his gait. If a shunt were not to be done, he recommended therapeutic spinal taps. Rowland agreed with Fahn; the two of us had once reported a favorable response to shunting in a patient with both cerebral arteriosclerosis and NPH.[1]

The situation was terrible, as might be expected when the patient is a most important person and a teacher with many medical disciples, all prominent but with no consensus about treatment. We knew that he had had multiple strokes and, therefore, had arteriosclerosis of blood vessels in the brain. What we did not know was whether treatment of hydrocephalus could help and how dangerous it would be to try that treatment. Merritt had said he would rather die than continue to be become progressively more impaired. He said this without evident depression; it was his usual cool clinical assessment.

Arrangements were therefore made to consult with Dr. Raymond Adams, chief of Neurology at Harvard, a former student of Merritt's, and the person who could be said to have "discovered" NPH as well as the favorable response to shunting. On December 14, 1978, Rowland went with Dr. and Mrs. Merritt to Adams and his equally illustrious specialist in cerebrovascular disease, C. Miller Fisher. There could not be a better pair of consultants anywhere in the world. The two of them interviewed and examined Merritt for about two hours. They found poor memory, as evident by inability to name presidents; bilateral upper motor neuron signs; shuffling gait with small steps; and imbalance with tendency to fall backward. Adams recommended a shunt on the basis of his experience with other patients who had both cerebrovascular disease and NPH. The chief neurosurgeon, William H. Sweet, agreed. He had probably placed more shunts than any other neurosurgeon anywhere.

Merritt was admitted to the Massachusetts General Hospital on December 14. Among other tests, Fisher noted improvement of gait after lumbar puncture and that the CSF protein content was only slightly high at 57 mg/100 ml, compared to an upper limit of normal at 45. The shunt was put into place on December 18, using a medium pressure Hakim valve to carry CSF from a ventricle in the brain to a chamber of the heart. (Hakim was coauthor with Adams on the first papers on NPH and had devised a valve that would help control the rate of CSF flow through the shunt, avoiding a sudden fall in CSF pressure.)

Merritt awoke after the operation on December 18 and seemed improved, but on December 20 he was drowsy and confused; CT showed that the ventricles were smaller, but movements of the left arm and leg were impaired. He became less responsive, but another CT on December 22 showed no change to explain the new problems. On December 23 he was given steroids, with some improvement, and the next day he sat in a chair. However, on Christmas day, December 25, he became unresponsive and never awoke. The CSF had abnormally low pressure, with a marked increase in protein content to 3300 mg/100 ml and 2000 red blood cells, indicative of bleeding into the CSF. CT showed a collection of subdural fluid beneath the skull; this was drained to relieve pressure on December 27. Then, the right femoral artery was occluded by an embolus on January 1, and the leg became gangrenous. Merritt was discharged on January 7 and transported to New York by ambulance. I rode with him. He was admitted to the Neurological Institute in New York for terminal care. He was totally unresponsive and died on January 8, 1979.

Autopsy findings were summarized by Dr. David Cowen, chief of the Division of Neuropathology.[2] He found two major disorders in the brain: vascular disease and Alzheimer disease. Severe atherosclerosis also affected coronary arteries, with recent thrombosis of the right coronary artery.

The pathology of Alzheimer disease is dominated by two changes: one is neurofibrillary degeneration, which was most prominent in the area of the temporal lobes called the "hippocampus"; the other is the formation of "senile plaques," which contain large amounts of amyloid, a substance made up of protein to which complex sugars are bound—these structures were numerous in all parts of the outermost layers (cortex) throughout the brain.

The vascular changes mostly affected "small arteries and arterioles at all levels of the nervous system and in all of the many regions sampled." The changes were "much more pronounced than is usually seen at this time of life," and hypertension was the likely cause. Although large cerebral vessels were spared, coronary arteries, the aorta, and the right femoral artery showed atherosclerotic pathology. Scars of ischemic infarcts were the result of occluded blood vessels failing to nourish immediately adjacent brain tissue and were found throughout the brain. Many lesions were judged to have been there for months or years; others seemed more recent in origin, perhaps weeks or days. There was no pathological evidence of Parkinson disease, and the ventricles were of normal size, with no postmortem evidence of changes that could have impaired absorption of the CSF. Cowen therefore considered the vascular disease sufficient to have caused the gait disorder. Alzheimer disease was responsible for the dementia and may have contributed to the gait disorder. Reciprocally, "vascular dementia" is currently a focus of research interest. It seems likely that both the vascular disease and Alzheimer changes brought on both the dementia and gait disorder.

Additional lesions were hemorrhages of recent onset, only days or weeks old; some were related to clots in cerebral veins, and other acute changes were likely part of the damage caused by postoperative decompression of the ventricles.

Cowen's note expressed skepticism about the diagnosis of NPH. However, the ventricles had been enlarged in life, as documented by CT; and clinical improvement followed removal of CSF by lumbar puncture. In life or even in the morgue, NPH is not readily diagnosed.

Notes

1. Earnest MP, Fahn S, Rowland LP, Gambetti P. Normal-pressure hydrocephalus with cerebrovascular disease. Trans Am Neurol Assoc 1972;97:268–270.
2. David Cowen died in 2001 at the age of 91. He had served at Columbia for 50 years and wrote 70 papers. He was a longtime associate of neuropathologist Abner Wolf, and they worked together to identify infantile toxoplasmosis, long before that condition was recognized as a secondary infection of AIDS. He was president of the American Association of Neuropathologists in 1962 and received a special award from that group in 1979 for "meritorious contributions to neuropathology." That same year Raymond D. Adams also received the same award.

Index